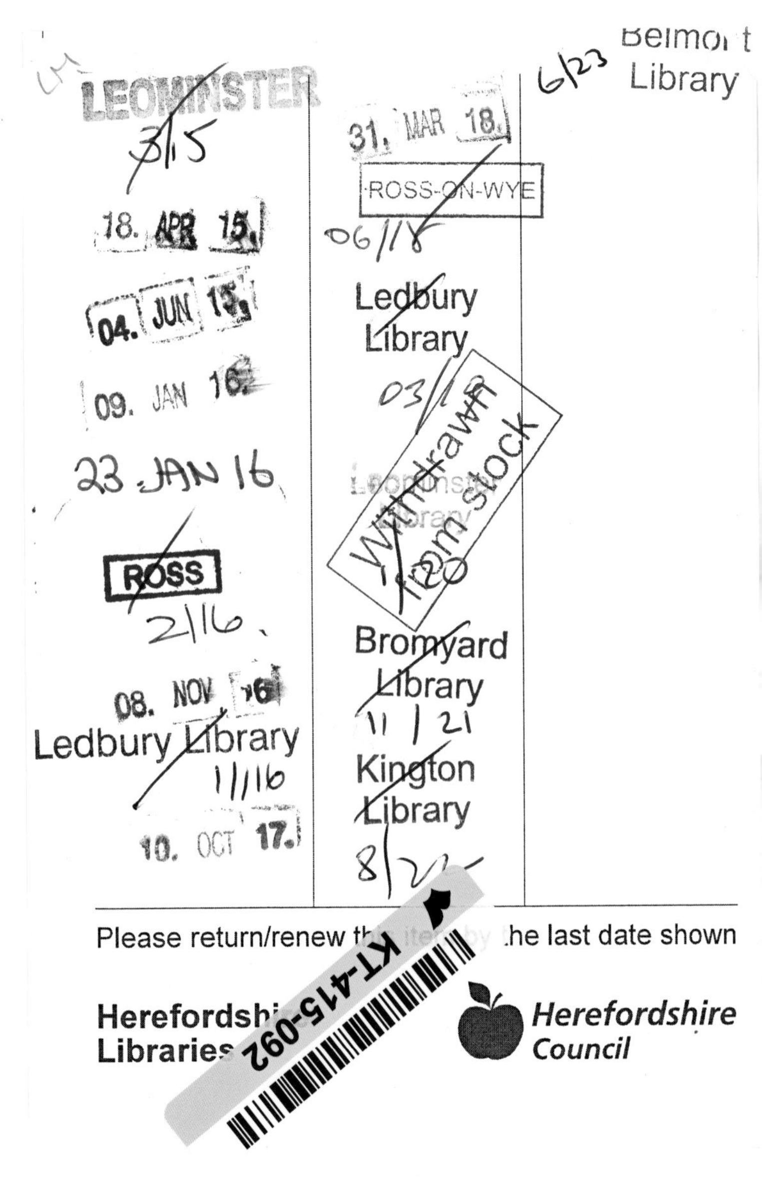

740006857055

Bill Oddie

UNPLUCKED

Bill Oddie
UNPLUCKED
Columns, blogs and musings

Bill Oddie

BLOOMSBURY
LONDON • NEW DELHI • NEW YORK • SYDNEY

Bloomsbury Natural History
An imprint of Bloomsbury Publishing Plc

50 Bedford Square
London
WC1B 3DP
UK

1385 Broadway
New York
NY 10018
USA

www.bloomsbury.com

First published 2015

A catalogue record for this book is available from the British Library.

Library of Congress Cataloguing-in-Publication data has been applied for.

ISBN (HB): 978-1-4729-1531-3
ISBN (ePub): 978-1-4729-1532-0

2 4 6 8 10 9 7 5 3 1

Typeset by Deanta Global Publishing Services, Chennai, India

Printed and bound in Great Britain by CPI Group (UK) Ltd, Croydon, CR0 4YY

Contents

Foreword 9

CLOSE TO HOME

Chapter 1: How to Be a Ludicrous Gardener 15

Chapter 2: Not in my Backyard 19

Chapter 3: Young People Today 22

WILD WORLD

Chapter 4: New-fangled Birding 27

Chapter 5: Climb Every Mountain 30

Chapter 6: Birds Online 34

Chapter 7: It's an Ill Wind 37

Chapter 8: Birding in Transit 42

Chapter 9: A Whale of a Time 48

Chapter 10: Galapagoing 53

BIRDS ON THE BOX

Chapter 11: They Couldn't Do That Now! 61

Chapter 12: Anthropothingy 65

Chapter 13: I Know That Tune 68

Chapter 14: Don't Look Now 70

INFINITE VARIETY

Chapter 15: Out for the Count 75

Chapter 16: Blowing in the Wind 77

Chapter 17: Puffin Billy 80

Chapter 18: Flying Tonight 85

Chapter 19: The Call of Nature 88

Chapter 20: Black and White 92

Chapter 21: Between the Devil and the BBC 95

Chapter 22: One That Got Away 101

Chapter 23: Invisible Bird 106

CURIOUSER AND CURIOUSER

Chapter 24: In the S**t 111

Chapter 25: Jewels in the Dark 114

Chapter 26: On the Fairway 117

Chapter 27: Birth of an Island 120

Chapter 28: Hearing Test 123

Chapter 29: Going Overboard 126

Chapter 30: Beware of the Cat 129

LET'S FACE IT

Chapter 31: Down on the Farm 135

Chapter 32: Protecting What from Whom? 138

Chapter 33: Meet us, Don't Eat us 141

Chapter 34: The Way it's Gone 146

Chapter 35: Harris Hawks 149

Chapter 36: First Mornings 152

Chapter 37: Going Bananas 155

Chapter 38: Crimes Against Nature 159

Chapter 39: Talking About your Generation 163

FALLING OFF A BLOG

Blog 1: News of the Wild 169

Blog 2: Frogs' Porn 174

Blog 3: Whodunnit? 182

Blog 4: Weather Report 189

Blog 5: Mine all Mine 194

Blog 6: Blow Me Down 199

Blog 7: Shall We Dance? 204

Blog 8: This Just in 209

WHO AM I?

Epilogue: Do they Mean Me? 217

Index 220

Foreword

Nearly every day people smile at me in the street. A delightfully high percentage say something like 'Hi Bill' or 'How are you doing, Bill?' or even 'Thanks for all the pleasure you have given me over the years.' It doesn't come much more flattering than that. I love it. However, there are a few people – just a few – whose approach is not quite so sensitive. They tend to stare at me for a bit and then ask: 'Didn't you used to be Bill Oddie?' I usually mutter: 'I still am'. But such folk are immune to irony. They continue: 'I thought you were. You used to be on the telly!' As if I might have forgotten. I thank them for reminding me. 'Yeah, with those two other blokes, you used to ride a bike.' The memories come flooding back. 'Yeah, you, Tim Brooke something and… the other one.' As it happens Graeme Garden is used to being called 'the other one' and he wouldn't be bothered. But I am. By remembering me only for my part in *The Goodies* 40 years ago, this bloke has dismissed – or maybe just missed – the product of half my working career! I could ask him what he thought of my wildlife programmes, but I can anticipate the answer: 'My mum likes those.' He doesn't need to add: 'Can't stand 'em myself,' though he may imply that by asking me to autograph a bus ticket while informing me: 'It's not for me.' I sign: 'Goody wishes to mum,' which hopefully will keep them both happy.

Truthfully and fortunately, the public enquiry I get most often these days is: 'When are we going to see you on telly again?' When indeed?! My only possible – and truthful – response is: 'It's not up to me. Write to the BBC. Please!' They may well already have done, but nothing's happened yet. But – just for the record – I am still Bill Oddie.

So, the next question has to be: what have I been doing for the last three or four years? Well, one of them was largely spent in and out of bed and psychiatric hospital, suffering from what was eventually diagnosed as bipolar disorder. Looking back now, I can reassess periods and incidents throughout my life when my behaviour fitted the basic pattern of manic depression swinging between being miserable and grumpy, or hyperactive and extrovert. The deep depressions are easy to label as a mental illness, but the manic and belligerent episodes get labelled simply as 'character'. Sometimes productive and creative, and sometimes a pain in the arse. Yep, that's me. Or at least, that's what I used to be. These days, people who should know – like my family – say that I am a much more amiable person. I am also out of work.

I won't go into all the details of my departure from *Springwatch* and *Autumnwatch*, partly because I have never been fully able to work them out. Suffice it to say that I believe that a manic phase of the bipolar was the reason that I was eventually told: 'We won't be asking you to do the next series.' It almost sounds like a judge teasing a contestant on *The X Factor*. 'We won't be asking you... We'll be telling you!' At which point I hug the BBC Head of Natural History! (That'll be the day.) No such luck. The words I was hearing were a euphemism for 'You're fired!' I almost wish that was what they'd said. I imagine it is much easier to be indignant and defiant when you are curtly told: 'You're fired.' If they'd said that I would probably have stomped out. Instead, I cried.

Frankly, I never did get a detailed explanation, but I have to acknowledge that during my final *Autumnwatch* on Brownsea

Island and even more so during location filming for a series called *Inspired by Nature*, I was ultra-critical, domineering, over-confident, impatient, etc. These are all things I have been on and off all my life, but in a state of bipolar-type mania, everything is heightened and more overpowering for those you work and live with. I am not saying that the BBC fired me because they realised that I was bipolar. They didn't know. And neither did I until the last month of the following year during which I was twice hospitalised for my own safety (a euphemism I'll leave you to decipher) and had to meet the crippling costs of private care, which didn't make me feel any cheerier! Ironically, it was when I was admitted to an NHS Crisis Centre that a doctor announced: 'I am putting you on lithium.' Only a week or two later, I heard one of my daughters ringing round with the glad tidings: 'He's back!'. Just in time for Christmas! By the new year I felt fine and I have been that way ever since. I can't claim that lithium is a guaranteed cure for bipolar but it works for some people. It has worked for me.

Right then, that's got that stuff out of the way. But the question remains: if I haven't been on the telly for a couple of years, what have I been doing? Once I was optimistic enough to accept that my recovery would be permanent, I made a decision to spend much more of my time supporting the organisations I admire and knew were contributing so much in the areas of nature conservation and animal welfare. I am happy to say that most of them (maybe all) were welcoming and appreciative. Over the past two years, you may have been aware of my sporadic appearances on the media on behalf of the Wildlife Trusts, the RSPB, RSPCA, Compassion in World Farming, The Humane Society, the League Against Cruel Sports, the International Fund for Animal Welfare (IFAW), the World Land Trust, Global Witness and others. I have tangled with ministers, appeared on *Newsnight* (though not with Paxman) and have accumulated a fine collection of security passes to the Commons and the Lords, though I have yet to appear on *Today in Parliament*! In tune with the zeitgeist I have become an ambassador (never without my Ferrero Rocher) for several NGOs (non-governmental organisations). In this capacity, I have visited Brazil, Argentina, South Carolina, Iceland, Guatemala, Zambia and Borneo.

During this time I have seen quite a lot of wildlife – much of it endangered. Whales, orangutans, elephants, and a fair number of new birds for my world list. However, wonderful though these creatures be, undoubtedly the most satisfying, enlightening, inspiring and enjoyable aspect has been meeting some terrific human beings. I don't mean to be ungracious about lots of lovely people I have worked with in the past, but – put it this way – people involved in the world of NGOs tend to be, well, nicer than those in broadcasting, and a *lot* nicer than those in politics. Here's a catchphrase for life: 'Doing good, does you good.'

The other thing I have been doing a lot more of is writing: articles, essays, blogs, tweets and so on. In this book I have hopefully collected some of the most interesting, contentious, amusing, etc. For me, there are two major attractions about writing or compiling a book. One is that I am allowed to get on with it, at least till the editor reads the final version and insists on lots of changes and rewrites (note to ed: please don't). The other attraction is freedom from a word count. Most magazine articles have to fit one page and so musn't be much more than 600 words or, for two pages, 1,200 words, and so on. Whatever the word count, it is never enough. Non-writers may assume the shorter the easier, but if you have any enthusiasm or imagination you need space to indulge it. I can honestly say that every single article I have ever written has been too long and I or the editor have had to cut it down. Sometimes I weep as a gem of a sentence or a joke has to go. This book is my chance to reinstate some of them. Many of these pieces are first drafts, uncut, the length that nature – and I – intended. So even if you think: 'I am sure I read this one in *BBC Wildlife,*' keep going, 'cos I might have had to cut the best bit, and now I have the opportunity to put it back. I might even have added a bit more.

CLOSE TO HOME

CHAPTER ONE

How to Be a Ludicrous Gardener

'It's getting ludicrous,' my wife told me. She was referring to my garden. She meant it as a criticism. I took it as a compliment. I am not sure that 'ludicrous' is the appropriate word, but I admit my garden is not conventional. It is not meant to be.

It isn't big – about half a tennis court. I have divided it into clearly demarcated areas, some of them inspired by my international travels. For example, there is the 'magic tree' – actually a lilac – festooned with the kind of shiny, glittery, swivelling things you can buy at a 'New Age' shop, augmented with strips of coloured paper and a few fake insects. Roadside trees in India are often decorated thus. Then there are the 'Inca Ruins', created from a few crusty old house bricks, some fragments of clay pottery, and one of those plaques with a smiley sun's face on it, which looks Inca-ish, even though it is actually from Camden Town Garden Centre. There is also a thriving thicket of tropical jungle, with ferns, palm trees, bits of bamboo and a couple of Buddhas. This I originally called 'Vietnam', even though I had never been there. I have recently renamed it 'Borneo'.

I have also excavated five small ponds, divided by equally small rockeries, which look almost natural. The same can't be said of 'Gnome Corner', the home of an ever-increasing colony of nearly 100 gnomes, some pleasingly jolly, others morose, but most of them undeniably tasteless. The whole effect is – what is the word? Imaginative? Quirky? Unconventional? Oh, OK. Ludicrous!

But what about the wildlife? To be honest, at first glance, one could be forgiven for assuming that my garden has been specially designed to deter birds. For a start, it is not exactly quiet. When a breeze blows, windmills whirr and wind chimes of all sizes tinkle and clang, like an under-rehearsed Balinese gamelan orchestra. Then there are the plastic predators. Lots of them. Many gardens have one. Usually a plastic heron, posing by the pond to deter real herons from gobbling up the goldfish. Have you noticed that it doesn't work? For a very good reason. If a flying heron looks down and sees what appears to be another heron crouched immobile over the water, it assumes that there are fish to be caught, and is most likely to join its chum, real or plastic. Wildfowlers use the same technique to lure ducks. They put out decoys. The fact is that fake birds are, if anything, likely to attract others of the same species, not scare them away. I have three herons and two egrets in my garden. Admittedly, I have only ever had one real heron, but then I haven't got any fish.

I also have several of those model hawks and owls that are meant to frighten birds away from airfields and valuable crops. So, do they clear my garden too? I conducted an experiment. I put a very realistic plastic Peregrine Falcon on my shed roof. This is a bird feared by many others, especially pigeons, which make up the staple Peregrine diet. I returned to the back room and lurked by the open door with my camcorder. Within 10 minutes I had video of three Woodpigeons pottering round the Peregrine as if it were one of their own. It was almost as if they knew perfectly well it was a fake.

To confirm my conviction that birds are not daft, my garden is now guarded by the Peregrine, a Kestrel, an Eagle Owl, a Tawny Owl and no fewer than five Little Owls. I have pictures of them with Great Tits, Blue Tits and Robins perched on their heads! I even once – in the cause of science – put out a totally realistic model cat (not a stuffed real one). Barely a minute later, that too had a couple of Wrens hopping on its back, and a young rat gnawing at its tail!

Then there are my mirrors. The generally touted advice is that you should not put a mirror in a garden, because small birds will attack their own reflection, thinking it is a rival, and they could hurt themselves by pecking at the glass. I have indeed witnessed that kind of behaviour, but only by Dunnocks, and though there was a fair amount of fluttering, pecking and poking, I am pretty sure there was no beak-breaking or head-bruising. I even solicited a second opinion by placing a vanity mirror on the bird table. It got no reaction at all, although I reckon one particularly vain Robin did have a bit of a primp. Anyway, my noisy, garish, model-infested, ludicrous garden has five large mirrors and a number of little ones or even shards of mirror glass, suspended on fences, nestling in rockeries and wedged in bushes. Thus the many psychedelic features are reflected and multiplied, as if in a giant kaleidoscope. There is even a large mirrorball, which on sunny days sends showers of sunbeams dancing over the lawn, and even on the ceiling of the back room.

Despite all of this, I have 58 species on my garden list. Admittedly, this is over 20 years, and a fair proportion of them were flying or circling overhead, including eight birds of prey: Kestrel, Sparrowhawk, Hobby, Red Kite, Buzzard, Honey Buzzard, Osprey and (real) Peregrine. Most birdwatchers count 'flyovers' on their garden lists – 'species seen in, over or from your garden' – except the RSPB. The rules for the annual Big Garden Birdwatch clearly state: 'Please record the highest number of bird species seen in your garden (not flying over).' Sounds a bit mean, doesn't it ? Well, no, it is all in the interest of facts, figures and fun. Anyway, it doesn't mean you can't look at anything in the air. Including the sunbeams. But one word of caution. If you do have mirrors in your garden, make sure you don't count every bird twice.

CHAPTER TWO

Not in my Backyard

My neighbours are not insensitive to nature. They think wildlife is fine. In its place. Its place, however, is 'somewhere else'. Not in, on or over their home.

We were recently gossiping in true neighbourly fashion, when we were distracted by a flypast of the 'Green Arrows', a small squadron of Ring-necked Parakeets. I smiled. My neighbour winced.

'You must agree, Bill, that really is a horrible noise?' I did not agree. 'They're going to cull them though,' enthused the lady of the house. I disillusioned her: 'No. They're culling Monk Parakeets. Ring-neckeds are definitely here to stay.' 'But that dreadful noise! It's like, it's like…'. She searched for an aptly offensive synonym. 'It's like a dentist's drill!' I was truly taken aback. Not by the vehemence of her outrage, but by the inappropriateness of the comparison. A dentist's drill? The call of the parakeet could perhaps be likened to the demonic screeching of a banshee, or maybe a

Jay with a loudhailer, but surely not a dentist's drill! I left them exaggeratedly covering their ears, so they didn't hear my parting mutter: 'Think yourself lucky you don't live in Australia.'

So my neighbours don't like nasty noises. Well, neither do I. Especially the most intrusive, ear-offending, teeth-tingling and unnecessary noise of modern times: the relentless roar of a leaf blower! And – here's an irony – from whose garden does this noise pollution emanate? My neighbour's!

It seems to me that there are people who don't mind watching a bit of nature, but are not so keen on living with it. The bloke on the other side loathes Woodpigeons. He once caused several panic calls to 999 by firing fusillades of gunshots from his garden terrace. When I enquired what the **** he was doing, he replied: 'It's the pigeons. They keep messing on the garden furniture. It's OK, they're blanks.' I know that Woodpigeons produce purple poo that quick-dries to the consistency of reinforced concrete, which can not only desecrate garden furniture, but also obliterate a windscreen, nevertheless, that is no excuse for sitting in your town-house garden blasting away with a shotgun, blanks or no blanks. As the police told him.

Of course, it is all a matter of attitude. Just as one man's weeds are another man's wildflowers, one man's pests are another man's 'pets'. Next door's wasps are likely to feel the fatal thwack of *The Sunday Telegraph*. Mine get coaxed into a wine glass with a postcard and are thrown to freedom. I found the velvet-furred family of baby rats that played hide and seek among my garden gnomes two years ago as cute as any mice. Next door started muttering about 'getting a man in'. I suspect he did. I haven't seen a rat since.

I am sure somebody did get a man in last year, when three fox cub corpses were found in various gardens, but I am certain that this slaughter was not perpetrated by the folks next door. They have become rather fond of our local fox, even though – or perhaps because – he is a bit of a scamp. One recent morning, I heard an incensed curse from over the garden fence: 'That bloody fox!'

'What's it done now?' I enquired with appropriate neighbourly concern.

'The damn thing's stolen my slippers!'

'He came in the house?'

'No. Yesterday evening, I left my slippers out on the patio, and he's had 'em!'

I had to chuckle. So did his wife. Eventually, so did he. 'Ah well,' he sighed philosophically, 'I'll just have to take it out on the leaf blower.'

'You do, and I'm going to get next door's shot gun!'

My threat was obliterated by the cacophony of the infernal machine. It also covered the startled squawking of half a dozen panicking parakeets that had been dangling on my peanuts.

As they swooped over my neighbour, he looked up, and I swear I read his lips: 'Mm. They are rather pretty. But what a horrible row!'

I should say so. He sounded worse than a dentist's drill.

CHAPTER THREE

Young People Today

Earlier this year the National Trust published a booklet called *Natural Childhood*, motivated by the assumption that most children these days don't have one! Indeed, the syndrome is so dire that it has been designated an adolescent illness: Nature Deficit Disorder, which – translated from the American – means 'not enough nature'.

The national newspapers were intrigued, especially by a list entitled: '50 things to do before you are 11¾', which could, I suspect, be retitled '50 things National Trust members used to do when they were kids'. No offence, but I can't imagine that many primary schoolchildren are National Trust members, or that they'd appreciate oblique Adrian Mole references. The list is clearly a nostalgia-fest, compiled by 'oldies', which could have been headed: 'What we used to do when we played out.' A phrase now long obsolete.

So what is on this pre-teen bucket list? Some suggestions are obvious enough. ***Climb a tree. Make a mud pie. Bury someone in the sand***. That one worries me a bit, especially in the absence of ***Dig them up again*** or ***Turn yourself in at the police station***. Nevertheless, as I perused further, I often nodded my approval, but I also noted the absence of any of the things that had connected *me* to nature when I was a little lad. Perhaps the National Trust decided not to condone them because they were illegal! Such as...

Scrumping apples. A euphemism indeed. *Stealing* apples more like. During 2011's riots people weren't saying: 'Hey, let's go and scrump some 42-inch TV sets!' Scrumping was fruit theft, but it lured the immature me into orchards and gardens, neither of which were familiar features of industrial Lancashire where I lived.

Bird egg collecting. Now illegal, and always reprehensible, but every schoolboy did it, and whenever I am asked how I got into birds and birding, songs, calls and habitats, I can only truthfully answer: 'By going egg collecting'.

Trespassing. When I were a lad, trespassing was unavoidable, simply because anywhere that was good for nature was also private. Notices told us that 'Trespassers Will Be Prosecuted'. Or to 'Keep Out'. Of course, we didn't, though getting in often involved danger or pain. Reservoirs were usually ringed by iron railings with spikes on top. Climbing them risked being skewered like a kebab. Sewage farms were fortified by high walls with broken glass on top. Farmland was more protected than a prison camp. The amount of barbed wire we had to climb over or under would have challenged Steve McQueen on his motorbike. Nor did we always make a Great Escape, and many's the time I felt the cuff of a farmer's horny hand, but it was a small price to pay to snaffle an egg from a Lapwing, Skylark or Yellowhammer, while honing nest-finding skills that Sherlock Holmes would have been proud of. So, I can't deny it, the thing that connected me to nature was crime!

But lurking behind my facetiousness, perhaps there is some deeper truth. A large part of 'playing out' was the thrill of dares and danger. The excited nervousness of possibly getting caught, or even prosecuted, whatever that was!

The fact is, we wanted to be naughty. All kids did, and all kids still do.

It is a fact that the National Trust booklet acknowledges. It could be subtitled 'How to let kids take acceptable risks'. However, will those risks be acceptable to health and safety? Or to paranoid parents? Or can they compete with the 'virtual' risks you can take in the latest version of Tomb Raider or whatever?

Changing the habits of children won't be easy, but there is an even tougher task. It is not just children who are suffering from Nature Deficit Disorder. We all are. So is the whole world.

WILD WORLD

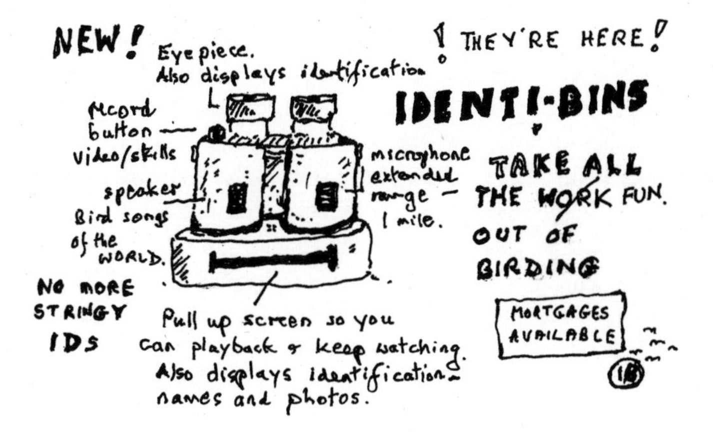

CHAPTER FOUR

New-fangled Birding

'When I were a lad, we didn't have binoculars. We had to make do with two toilet rolls and an elastic band.'

'And if you wanted a telescope, you carved one out of a marrow.'

'And there were no bird books. None at all. I tell a lie, there was one. Only it didn't have any birds in it. It didn't have anything in it. No pictures. No words. No point! We never saw anything. Nothing at all. But that didn't stop us going birdwatching. Every weekend, we'd ring our friends and tell them where we'd seen nothing, so they could come and see nothing with us. Things have changed since then.'

They certainly have. What's more, I am old enough to have witnessed – and benefited – from the many advances in birding equipment and techniques. I have seen telescopes turn from burnished brass into grey rubbery plastic stuff and shrink from the size of a blunderbuss to a small vacuum flask, thus evolving into 'spotting scopes'. Binoculars have basically retained their shape, and presumably always will – unless we evolve a third eye – but they are lighter, optically impeccable and bank-breakingly expensive. Bird books are arguably out of control, and indeed there may well now be more books than birds. In fact a cull may be necessary to make way for the proliferating DVDs, websites, apps and so on. The advances in communications have been entirely beneficial. Time was when birdwatchers rang each other and relied on 'the grapevine' for news. This became focused by Birdline's Rare Bird Alert, which could be consulted on a new-fangled 'mobile', which begat the 'pager', and has now evolved into the 'smartphone'. Don't you just hate that word 'smart'? It sounds so smug. Or is it me being intimidated by technology?

The truth is that, having been out of the loop (and into the 'loopy'?) for a year or two, I have only recently re-emerged into the current birding scene. I chose to dive in at the deep end by going on a trip to Guatemala, where the birds were almost totally unfamiliar, and the forests dauntingly dense. My first day out, I floundered.

'What's that?'

'Where?'

'There.'

'It's flown.'

'It's back.'

'Where?'

'There.'

'Gone again'.

'Lost it. Aaagh!'

I have rarely felt so inadequate. Nor so envious of my companions, who were mostly calmly and confidently calling out the sort of names you only get in South America: 'Scaly-throated Leaftosser.' (honestly!)

'Blue-crowned Chlorophonia.' (is that a bird or a disease?)

'Northern Bentbill!'

'Oh, come on! It can't be.' It is. Just flown out of the palm tree. And ... 'What's *that*?' This time there was no 'call'. I realised that I was about to get my first demonstration of birding in the digital age. This is how it works.

The first requirement is physical. You need to be strong enough to carry binoculars, a telescope on a sturdy tripod, hooked over one shoulder, and a huge backpack, containing all manner of digital gear – soon to be revealed – and a laser pen. Yes, one of those things some twerp tried to blind Ronaldo with, or dazzle Andy Murray. I had regarded it as a weapon, until I went birding in Guatemala.

More than once I watched the routine spring into action. It went like this.

1: someone spots a little brown bird. But 2: no one knows what it is. While 3: I can't even see it. So 4: somebody points the laser pen, and a red or green dot flits up and down a huge tree until it hovers and stops. Pen man announces 5: 'The bird is 6 inches above the light.' Which, 6, it is. But what is it? Then 7: strong man steps forward, with digital camera fitted with a lens as big as a bazooka. The bird is tiny and nearly half a mile away, and yet the camera clicks and whirrs. Next 8: we all peer at the screen. A brown speck is just about visible, until 9: the digital zoom is activated, and the image magnifies so rapidly it is as if the bird is charging straight at us. Magnified to ultra-close-up, we can now see every stripe, streak and feather margin. The cameraman is even able to calculate the length of the wing! Until 10, an identification is mooted: 'Probably Yellow-bellied Flycatcher'. By now, at least three smartphone screens are touch-scrolling through Flycatchers of Central America, *while a nearby iPod broadcasts a wispy snatch of virtual birdsong, at which the non-virtual bird flies closer and joins in, and 'probably' becomes 'definitely'.*

Yellow-bellied Flycatcher. Tick.

So, is it progress? Or is it 'cheating'? Or do you miss the days of toilet rolls and marrow? I do.

CHAPTER FIVE

Climb Every Mountain

Any birdwatcher who keeps a world list is bound to have visited a few volcanoes. In order to see a Volcano Junco, you must go to Costa Rica and ascend to the edge of the crater of the Poás Volcano. It's not too arduous if you do it in a charabanc. The junco isn't spectacular: it looks rather like a smoked sparrow and is lava-coloured. So are a lot of lava-loving birds. On the volcanic Galapagos, the Lava Heron is black. So is the Lava Gull. (A seagull that isn't white! Doesn't seem right does it?)

Of course, these species are examples of a well-known natural law: that creatures are often the same colour as their surroundings. Mind you, by that token, rainforest birds should all be green, and wet. Some of them are. However, others are dazzlingly and almost luminously multicoloured, even some of those that live on volcanoes. But then not all volcanoes are covered in lava.

About a year ago, I joined a small party of British, American and Dutch journalists who had been invited to sample some of the birds of Guatemala. We were guests of Guatemala Nature Tours. We started with a couple of days of sociability and some gentle lowland birding, including a cruise across one of the calmest and bluest lakes I have ever seen, made all the more stunning by the reflections and backdrop of not just one but several volcanoes. Every now and then we were distracted by a distant rumbling. Thunder? Quarry blasting? It was coming from high in the mountains. Our binoculars revealed plumes of smoke and at least two of us swore we saw the earth move.

We were, however, assured that an eruption was not imminent. This was comforting news, since we had been told that our post-breakfast walk was going to be 'up the volcano trail'. It was at this point that I beckoned to our tour leader. I explained that I had spent a fair chunk of the past year in and out of hospital, and that the only walking I had done for several months was a casual stroll on Parliament Hill, which is quite steep but hardly competes with a volcano. Another member of our group sensed my trepidation and expressed concern about his 'dodgy knees', while a third asked that his advancing years be taken into account, at which I muttered that my years were further advanced than his. The tour lady assured us that we would all be fine. 'It is a good trail. Up through the forest. We will have a little rest at the lookout, then carry on to the crater if you want. Or we can just keep birding.'

'How long will it take us?' I asked. She hesitated. Was it because she didn't know or because she didn't want to tell us? 'About three hours I think.'

'There and back?'

'Three hours each way.'

'So six hours!'

'Maybe. Yes.'

It took us 12! Six up. Six down. Twelve hours! Well that's how long I took. The youngest and fittest took maybe about eight. Dodgy-knee man took nearer 10, while Mr Advancing Years apparently gave up at the lookout.

Oh yes, while we are at the lookout, what did we look out at? Trees. Miles and miles of trees. The same trees that grew alongside and over the trail so that the sun rarely broke through. The same trees whose tangled roots snaked upwards through and across the path and grabbed our ankles and tripped us up. The trees that hemmed us in on all sides and made sure the view from the lookout was the only one we got. That's the thing about volcanoes, once you are on 'em, you can't see 'em. Snow-capped peaks can only be viewed from a distance.

We trudged upwards and then downwards. I fell further and further behind the main party. A guide was assigned to stay with me, to catch me if I collapsed and, if necessary, to carry me. It wasn't necessary – quite – but, at least a dozen times coming down, my legs turned to jelly and gave way, so that I slid on my back for several yards, always yelling: 'It's OK! It's OK!' to reassure my sherpa that I hadn't had a heart attack.

Then it began to get dark. Any recognisable friendly voices had long since faded way ahead and below us. I had visions of my so-called friends piling into our minibus and driving off to quaff cold beers at a local bar. Meanwhile, somewhere on the mountain, there was me and my guide. Babes in the wood. We were enveloped in total blackness, apart from the feeble beam of a small torch, which cast shadows and made the forest even spookier. As I stopped for yet another rest, my wheezing and heavy breathing added to the soundtrack until it subsided into silence. If it had been a movie, there would have been a pause, then an ominous distant rumble, exploding into a mighty crash. And a panicked voice announcing: 'The volcano! It's erupting!' But it wasn't.

There was, however, a more incongruous sound approaching: Michael Jackson singing 'Thriller'! Then, a manly duo joined in the chorus in what I suspect were meant to be zombie voices. Not me and my guide. Two armed policemen who had been sent to make sure we came to no harm. I was relieved. Both by their presence and by the fact that it had only just struck me that – like most Central American countries – Guatemala has not long been free of bandits and guerrillas. Unfortunately, despite warnings, European birders are notorious for wandering off into the woods in pursuit of some local speciality.

Talking of which, you may be wondering what birds we saw up the volcano. In all honesty, not a lot. There were several species of wintering North American warblers, and the occasional glimpse of a woodcreeper, but frankly we'd had better views in the lowlands where the canopy wasn't quite as dense. There was, however, one specific reason we needed to get up to high altitude. To search for a quintessential ticking target, the Highland Guan, which sounds like a Scottish folk dance, but looks rather like a tree-dwelling turkey. It has red feet and a red wattle, but is otherwise, well, lava-coloured. It is also big. Worth a 12-hour hike?

I wouldn't know. I didn't see it. None of us did.

PS. It needn't happen to you!

Guatemala is a fascinating country, with stunning scenery and terrific wildlife. Just make sure you check the itinerary, and don't try and do too much. Mind you, if your main concern is the length of your checklist, choose a specialist travel company, and go for it.

CHAPTER SIX

Birds Online

In the olden days, nearly every year I would spend a week or so during the migration season on the tiny Shetland island of Out Skerries. There were no trees, no hedges and no electricity, which meant of course that there were no telegraph poles and no wires. Skerries was therefore in effect a perchless zone. Small birds had to make do with drystone walls and a few crofters' roofs. Then came a historic day. The generator chugged into silence, an underwater cable brought energy from the mainland, and up went the poles and the power lines. Most of the islanders were pleased, a few protested, but I was delighted. I was also excited, as I anticipated which species would be the first bird on the wire.

I predicted Wheatear, expected Meadow Pipit, but was thrilled when it was a Cuckoo, a bird we frequently hear but rarely see, unless it is perched out in the open – for instance, on telegraph wires. I am sure most birdwatchers have wondered:

'What would we do without wires?' More to the point: 'What would the birds do?'

Most 'accidental' man-made perches have their natural equivalent. A TV aerial is an iron treetop. A chimney stack is a brick tree stump. Both are ideal song perches. Likewise, drystone walls are rocky hedges, and dangling ropes could be lianas. But where in nature will you find a long length of highly strung horizontal cable, thin enough for a bird to grip, and with an unimpeded view, affording safety from predators and a lookout for prey, not to mention a symmetrical gathering point for the flock whose members like to line up together? Is there any image more iconic of migration than newly arrived or ready-to-depart Swallows dotted along the wires, like musical notes on the stave, or clothes-pegs on a washing line?

Poles and power lines are also much favoured by the 'perch and pounce' predators, principally birds of prey, such as shrikes and kestrels, with talons small enough to grip, and tails long enough to balance. Peregrines and buzzards pose on posts, often so still that they seem to be growing out of the top, or maybe carved like a totem, until they suddenly launch themselves either skywards, or downwards onto a rambling rodent. The most regal raptors claim the most majestic thrones. There is an area in Israel where a veritable forest of giant pylons undeniably constitutes one of the grossest blots on the landscape you will ever see, but it is arguably also the best place in the country for seeing a selection of large falcons and even larger eagles.

Poles and pylons also provide nest sites for a few species: Ospreys, storks, weavers and woodpeckers (as long as the pole is wooden, not steel!). Monk Parakeets build a ramshackle colonial nest the size of a small car and are capable of plunging a whole neighbourhood into blackout.

Talking of which, how come that the birds don't get electrocuted? Small birds are unlikely to touch two wires or the ground, and are therefore generally safe. However, the larger the wingspan the more the risk of double contact and electrocution, and many birds do die, especially raptors. Nevertheless, when power lines hit the headlines it is usually because of a bird strike, often by geese or swans. Sometimes the location of the pylons is criticised, but I admit I do have to question the steering competence of some Mute Swans, having recently watched a fully fledged free-flying bird flounder and flap across Hampstead Pond, take off and crash straight into a wooden fence!

So, poles and power lines, call them an eyesore or an obstacle, but I'm willing to bet most birds are grateful for them, and so are birdwatchers. Here are a couple of challenges: what is the least likely bird you have ever seen perching on a wire? (It has to hold steady for at least a minute, and no wing flapping.) Mine was a Curlew! And what's your 'most species line up' along a wire between two poles? My best was Portugal last August. Bee-eater, Woodchat Shrike, Southern Grey Shrike, Hoopoe, Roller, Red-rumped Swallow and Azure-winged Magpie, with a Little Owl on the post.

Electric!

Little old men in dinner jackets OK, but we do NOT look like NUNS!
and don't speak with your beak full....

CHAPTER SEVEN

It's an Ill Wind

Is weather wildlife? No, but weather is nature. And weather affects wildlife, especially wild weather. Not that we get really wild weather in Britain. We get rain, we get floods, we get cold spells and we are told we get droughts. Very occasionally, we get strong winds but, own up, you have to go back to the Great Storm (as history has recorded it) of 1987 to recall truly photogenic damage. I was filming an educational video at the Tower of London, but arrived to find the entrance blocked and continuity ruined, because what the previous day had been a grassy courtyard was completely obliterated by the massive tree that had stood proudly in the middle of it, but now lay sprawled and battered like a heavyweight boxer out cold on the canvas. These days, I would have snapped it with my iPhone and had it on Twitter in five seconds, where it would have joined thousands of posts depicting the aftermath of that uniquely stormy night, but back in that primitive era, we had to wait till the evening news before we could marvel or panic at the extent of the damage. Cars overturned, roofs blown away, whole woods flattened and wildlife – especially birds – compelled to go wherever the wind carried them.

Michael Fish may not have predicted the hurricane, but birdwatchers certainly anticipated its effect. The storm had gathered up its strength in the Bay of Biscay. The date was 18 October. So what seabirds would be in Biscay in October? Leach's Petrels, maybe shearwaters and surely Sabine's Gulls. The latter is a birders' favourite. Rather dainty and handsome (for a gull) and rare enough to be desired but not so rare that you don't stand a chance of ever seeing one, or even a handful, especially if you are on the west coast in a strong westerly which hampers the speed of their progress to wintering areas further south. On 18 October 1987, the wind in Biscay was so powerful it literally swept up fistfuls of Sabine's and hurled them north-east to the coast of Britain. Birders welcomed their arrival, and were doubly delighted when they started seeing birds – not down in Cornwall and Ireland – but in the middle of London and the Home Counties. All day, the mobiles and pagers trembled with new reports: Sabine's on the Thames, in parks, gravel pits and reservoirs. A record-breaking invasion never to be repeated, and it hasn't been. Yet.

However, in North America this sort of thing happens every year. Over there, extreme weather is accorded the status of an ancient god. It is to be feared and revered but it also has its fans and followers. There are clubs devoted to tracking down tornadoes and filming them at dangerously close range. Twister twitchers indeed. The TV weather channel – yes, showing non-stop weather and nothing else – mixes warnings, facts and forecasts with viewers' videos, which are viewed not with trepidation, but with almost childlike excitement. 'Oh boy, look at the tail on that! Wow, that is awesome! That's gotta be today's top twister. Filmed by Robert K. Cheeseburger, in Georgia. Way to go, Rob!'

Only once have I myself been close to a mighty wind. I was in New Jersey at Cape May, which is at the tip of a peninsula with the mouth of the Delaware River on one side and the Atlantic on the other. Hurricane Floyd had just finished battering the Bahamas and was now careering through the Carolinas towards Washington. It would then probably hit New Jersey. Hurricanes are unpredictable, but in the States they don't take any chances. As I drove down to Cape May I had already noticed permanent road signs indicating 'Evacuation Route'. They all pointed due north. Unsurprising, since the other three directions all led to the ocean! I was not aware of any panic on the streets of the rather sedate colonial-style resort of Cape May, despite the fact that the State Governor was on every TV screen declaring an official 'state of emergency'. It struck me that it was probably to avoid litigation. 'I can't force you to leave, but don't say I didn't warn you. And I hope your insurance companies have got that.'

The next morning was wet but not a deluge; the wind was strong but not lethal. Clearly the worst of Floyd had passed in the night. But what had he left behind? He'd been born not in Biscay but in the Bahamas. It was quite possible that he had scooped up some tropical birds and whisked them north up the east coast, even as far as Cape May, where binoculars would be palpitating with anticipation. A species that may be common in the Caribbean could be a celebrity in New Jersey. Somewhere a welcoming committee would be waiting, but where?

I rang the local birdline and was directed to a dilapidated beach café. I drove there immediately, leaving a splashy wake down the slightly flooded streets. Thanks to the admirable American style of having a numbered street sign on every intersection – why can't we do that? – I was soon skidding to a halt in the café car park. Clearly Floyd had been a bit rough with the café. What was probably an already flimsy building now looked on the brink of being totally blown away. Neverthless on the rickety forecourt were 20 or so blokes with binoculars and telescopes, staring out to sea and now and then yelling out adjectives. 'Sooty!' In the UK it would have been Sooty Shearwater, here it was Sooty Tern. Followed by 'Bridled', at home 'bridled' Guillemot, here Bridled Tern. Two species of 'tropical' terns abducted from the Caribbean. Both New Jersey rarities.

The locals were thrilled and I was pleased for them, but vicarious pleasure is not in a birder's nature. The truth was that I was getting more and more miffed because I couldn't see very well. Such is the penalty of being not much taller than a hobbit. These were nice guys, but nobody was about to let me push into the front row. However, just behind the café, was a grim grey building with a first-floor walkway, somewhat reminiscent of

Bates Motel in *Psycho*. 'Wouldn't it be better if we could get up there?' I asked. 'We'd be higher and drier. Can we get in there?' My words were greeted as if they were sacrilege, which indeed they were. 'That's a convent.'

'What? Like a convent school?'

'No, like a nunnery.'

'Oh. But are the nuns actually there?' I reasoned. 'They could've been evacuated. Or be away on a pilgrimage or something.'

'Lead us not into temptation,' somebody quipped. While someone else stuck his fingers in his ears to block out my wicked words. A third – continuing the frivolous blasphemy – raised his hands to the heavens and asked the Lord to give us a sign. At which moment, the café roof blew away, and one of the walls collapsed. It helped our decision. 'Aw, what the hell! Let's do it!'

Maybe the mention of hell unnerved me. Or the crucifixes or the Latin scriptures, not to mention the signs saying 'Strictly Private' and 'No Access'. At least two of us couldn't resist muttering 'forgive us our trespasses,' as we scurried up the iron staircase, our boots clanging like gongs on the metal steps. 'Are you sure the nuns aren't here?' I whispered. Even as I spoke, two black and white figures appeared in the courtyard. Identification was instant. Black and white Razorbill? Velvet Scoter? Black and white… Nuns! They didn't scream or yell at us, but neither did they ask if we had had any 'tropical overshoots'. Instead, they popped back out of sight. I for one was relieved. It was almost as if God had given us his blessing. He hadn't.

Ten minutes later three cops arrived. Proper US cops with gun belts and star-shaped badges and many-cornered caps, and – most unnerving of all – truncheons at the ready. 'What the hell are you guys up to?' drawled the chief officer, and then answered his own question. 'Goddam birders. Every time there is a storm I have to throw you guys out of here.' Oh my lord, I thought, they've got previous. I instantly decided to take the rap by playing the bemused Englishman. Which I was. 'It's my fault, officer. You see, this is my first hurricane. I suggested we came up here 'cos it's a much better view.'

'I am sure it is, but it freaks out the ladies. Do me a favour fellas, stay in the beach café, OK?'

'But it's nearly washed away,' I argued. The officer glared at me, just like my Latin teacher used to when he was about to say: 'Oddie, one more word out of you and you are in detention.' Only in this case it could be: 'You are in jail.' As we dolefully filed down the stairs, it struck me that if you are knowingly going to antagonise anyone you could hardly choose more intimidating foes than policemen and nuns!

We closed the iron gate with the big padlock and 'Keep Out' sign, with a line of nuns peeping over the balustrade like Puffins over a ledge. As the officer revved up his motorbike, he shouted over to us: 'Hey, fellas, one more thing.' We sighed a silent 'what now?' 'So, tell me, did Floyd blow in any good birds?' Recognising the opportunity for petty revenge, we collectively chorused. 'No!' In fact, in the brief time we'd been on the balcony we'd had 40 Sooties and 10 Bridleds, but we weren't going to tell him. Mind you, if any of the sisters were seawatchers, with a view like that, they must've seen even more. Gripped off by a nun! Well, there's a first.

CHAPTER EIGHT

Birding in Transit

Most birdwatchers keep lists of how many species they have seen – on their local patch, in their garden, in the current year, in their life, in Britain, in Europe, in the world. If they travel abroad in search of ticks (new birds) they probably keep a trip list. They may well also keep 'transport' lists: a train list, a bus list, possibly a bike list (you are allowed to dismount but not to abandon your bike). The long-distance traveller may even keep an international airport list of birds seen in transit.

As soon as the pilot has announced: 'We shall be landing in a few minutes,' you should start peering through the windows to claim the cachet of spotting the first bird of the trip. Its identity will vary from country to country and continent to continent, but not perhaps as much as one might assume. Whatever the latitude or longitude it will probably be something white: a gull, a gannet or a tropicbird, if the descent is over the sea. If it is inland, it is quite likely to be Cattle Egrets. My best airport white bird was a Snowy Owl in Canada. And no, it wasn't one of those fake owls they put on airport roofs to scare other birds away. They are usually plastic Eagle Owls and, incidentally, they scare nothing. Birds aren't daft.

Birdwatching during landing is rarely productive. The plane is travelling too fast and bouncing too much, and even if you do glimpse something, you are strapped in and can't move, let alone hold your binoculars steady. Once the plane has landed and begun taxiing your chances improve. The majority of the world's airfields are indeed sited on fields, or at least have large grassy areas. A lot of it may be closely mown, but there are often rough patches perfect for birds that like grasslands. Larks and pipits aren't easily named when you are still strapped in your seat belt but you can have fun guessing and anticipating. There is always a chance of a harrier. If it is a male, you may be able to name the species before you are allowed to unfasten, stand up and try to elbow your way towards the exit. If you are in transit nowadays chances are you will have to disembark, be stuffed into a motorised cattle truck, and be chivvied into the terminal transit lounge, where the only windows look out on parked planes, luggage trollies and concrete. At least there is the consolation of a few species for your airport list. Almost wherever you are in the world, you can tick off Common Mynas, House Crows and House Sparrows, which are probably also thriving in the nearby city. Why, then, have House Sparrows almost gone from London? Is Regent's Park really less salubrious than an international airport? And how come the local swallows and swifts choose to nest under steel beams on concrete hangers, constantly deafened by jet engines and breathing in toxic fumes? At least they have the airport grassland just round the corner, where they can skim around feeding on bugs and butterflies and drinking from the ditches, where frustrated birders can't see them. It was not always thus.

Sadly, nowadays, high-level security means that birdwatching at airports is likely to be limited. Gone are those delightful days when we were offered the choice of spending transit time in the terminal or to stay on board and even be invited to stand at the top of the steps at the open exit door, where you could get a breath of fresh air, and brazenly scan across the grassland and sort out those larks and pipits. It was in such circumstances in Singapore that I spotted a small flock of Oriental Plovers, which I was able to point out to a rather attractive air hostess. I even erected my tripod and let her look through my scope. It wouldn't happen nowadays.

Mind you, I suppose I got a foretaste of the future tension we have to accept these days back in 1990, when I was returning from a trip to East Africa. It all seemed almost unreal. We had spent two nights camping in tiny tents at the bottom of Tanzania's Ngorongoro Crater, where a nocturnal excursion to the canvas-clad loo involved tiptoeing past African Buffaloes and dodging the Lions that were hopefully now satiated by dining on the antelope we had watched them chase down that very afternoon. A guard with a gun accompanied me, which was comforting, albeit a little inhibiting. Two days later, I was revelling in the luxury and privacy of an en-suite bathroom in a three-star lakeside hotel with an African moon rising above a backdrop of the giant snowy cone that is Mount Kilimanjaro. Almost unreal. So was what happened next.

It was the dead of night and I was deep in sleep. Suddenly, I was woken by the sound of engines revving, vehicles manoeuvring and someone hammering on every door in the corridor and yelling out words I have never heard before or since: 'Wake up! Your plane is going early!' Surely not. Planes don't go early. Nothing goes early. Someone insisted that the night porter phone the airport. He said he had but the phone was dead. So how did he know the plane was early? One of our group demanded to have a go. The night porter handed over the phone while muttering 'out of order'. It was. He muttered something else in Swahili, which probably translated as 'bloody suspicious tourists'. He repeated his original threat: 'Your coach is leaving,' plus more Swahili, no doubt 'or don't you believe that either?' He almost pushed us outside. 'Quickly. Hurry, quickly.' 'Is there any other way of hurrying?' I quipped, as we were herded out and onto our vehicle, not even being given time to snap the sunrise over Kilimanjaro. It was as if they couldn't wait to get rid of us. Bleary-eyed, unwashed and slightly trembly, I sensed a slight air of panic, or was it foreboding?

In fact, the next part of our journey home went calmly, except that I did wonder why we were returning to the UK via Addis Ababa in Ethiopia, a land wracked by widespread starvation and poverty but also, at that time, civil unrest and armed uprising. Being at an airport at such times is not relaxing, the only consolation being that with any luck it means that it won't be long before you can flee the country. Once again I and my fellow passengers were accompanied by a man with a gun, who frankly lacked the protective demeanour of the loo guard in the Ngorongoro. This man looked scared. He herded us into the main hall and left us to stare at the departures board. There were half a dozen international flights listed. On all five, passengers were requested to wait in the lounge. The uncomfortable feeling changed to concern, as the display changed to 'delayed', which disturbingly became 'delayed indefinitely'. And after an hour or so, a positively scary 'cancelled'. There was no explanation. No announcement about what was or wasn't happening. There was, however, a multilingual rumour going round that all planes had been commandeered by the Ethiopian military to transport troops to fight rebels in the north. It seemed that we could be in Ethiopia for some time.

Which wasn't such a bad thing since Ethiopia is a terrific country for birds, including several much-prized endemics. Three of us were clutching our binoculars and scopes – birders never check in their optics – and we had already noticed that the airport grounds were promisingly green with grass, bushes and trees. And there was even the occasional tantalising flick or flutter. There were birds. We needed to get at them. The only exit involved large glass doors through which a few people were coming and going. Some were armed, others were in uniform. Non-military personnel were having to show ID, tickets or passports. Most were being allowed through. Would we be? And if we were would we be allowed back in? Do we risk it? Of course. Most birders walk a thin line between determination and recklessness.

A passport, a ticket and a smile got us through. Within seconds we had scampered across a car park with a larger population of House Sparrows than London. Then we noticed one we didn't recognise. Some kind of thrush. We were soon engrossed in scribbling descriptions and sketches, leafing through a Tanzanian field guide, which the mystery species probably wouldn't be in, or taking photos. This was probably our big mistake. Not one, but two, and then three men with guns scampered towards us, yelling more menacingly than the night porter. I say 'men', but none of them looked more than 13 or 14. Their rifles looked older, but no doubt worked. The fact that the boys' fingers were trembling on the triggers, their foreheads sweating and their voices yelling almost hysterically was evidence that they considered us suspicious if not hostile. We couldn't blame them. Unfortunately, birding optics look like surveillance gear at best, at worst they could be mistaken for firearms. Wearing camouflaged jackets doesn't help. The time-honoured birdwatchers' defence tactic when under suspicion is to show the soldier, policeman or enraged landowner a field guide, show him a few pictures and mime flapping, while repeating 'birds', and maybe even doing a few bird impressions. This routine works best in countries where the concept of watching birds for pleasure is not totally unknown. It did not impress teenage Ethiopian soldiers, whose nerves were already in shreds. We allowed them to march us at bayonet point back into the terminal.

We were now burdened with a double whammy. The planes were still 'cancelled', and even worse we knew there were new birds out there – maybe even once-in-a-lifetime endemics – but we couldn't get at them. So near and yet so far. I decided to console myself with an unescorted visit to the loo. Within seconds, I came rushing out and beckoned my two comrades to follow me back into the cubicle. Perhaps, surprisingly, they willingly did so. What any onlookers thought I can't imagine. Well, I can, but I'd rather not. We closed the door and clicked on 'engaged'. Tim, the tallest by far, squeezed between us and examined the window. Yes, a window. Quite a small window, and quite high up. It had frosted glass but it had been left slightly open for ventilation. Tim pushed it fully open, thus creating a luxury birdwatching hide, complete with uninterrupted view and built-in conveniences.

The only design fault was that the window was so high that only very tall people like Tim could look through it. Small people like me could stand on tiptoe and still barely reach the sill. All I could see was a sliver of blue sky. I could, however, hear Tim's commentary. 'Pigeons. Er, White-collared Pigeon. Endemic, I think. And Brown-rumped Seedeaters. Oh, wait, yes, definitely Wattled Ibises. That is an endemic! Can you two see all right?'

Imagine the scene. A toilet cubicle made for one but accommodating three. Three blokes crammed in like contortionists and lumbered with large backpacks, tripods, telescopes and cameras. There is a very tall man who is able to peer through the little window, a middle-sized one who can just see out if he stands on tiptoe, and a little one – me – with no choice but to clamber up to stand on the rim of the lidless toilet bowl, steadying myself by grabbing the cistern and risking pulling it off the wall and, on several occasions, when I lost balance, grasping at the chain, which saved me from slipping down the pan, but of course kept flushing. The scene reminded me of student days and 'How many people can you get into a mini?' or the Marx Brothers' routine with more and more people cramming into a cupboard. We were in there for maybe an hour. By the time we emerged, the 'cancelleds' had been replaced by boarding times and gate numbers, while the story about the military taking the planes had been dismissed as 'Chinese whispers'. Better still, I had added a dozen species to my international airport list.

That was 20 odd years ago. Nowadays, Ethiopia is a prime destination for birdwatchers. No doubt there are many books and websites advising 'where to watch birds in Ethiopia'. I do hope they include the gentlemen's toilet at Addis Ababa airport.

CHAPTER NINE

A Whale of a Time

I have now been on three whale-watching trips in – or should that be from? – Iceland. The first one was quite a few years ago, out of Húsavík in the north. We saw a couple of porpoises and a distant dolphin, but no whales whatsoever. This struck me as worryingly ironic. First, because the intention was to expand whale-watching as a tourist attraction. Second, because it was rumoured that the Icelandic whaling industry – dormant during a moratorium – had announced that it intended to resume its activities. It was a contradiction that to me seemed almost laughable, with only the small consolation that both whale-watching and whaling are a bit of an anti-climax if there are no whales!

In subsequent years, I have been on enough 'pelagics' in various parts of the world to realise that the fact that whales are not visible doesn't mean they aren't there. Any cetacean seeker will tell you there will be bad trips and there will be average ones, and now and again there will be a real mind-blower.

The second Icelandic trip, in August 2012, was not mind-blowing, but it was fascinating meeting and marvelling at the work of IFAW scientists monitoring the impact of the satisfyingly escalating whale-watching business, and meeting a splendidly cosmopolitan team of young volunteers who were intercepting tourists and appealing to them to 'Greet us. Don't eat us' – a plea on behalf of the whales, not the volunteers. A charmingly petite Polish girl had even disguised herself in a less than life-size whale costume, which was extremely hot, heavy and had no proper eyeholes. She could all too easily have toppled off the quayside and into the water, especially if 'helped' by anyone protesting about protesters, as it were. 'Try anything' tourists intent on sampling local specialities like whale steak can get belligerent if their dining plans – or even their morality – is challenged. IFAW's lobbying is peaceful, but the response may not be. Fortunately, on this trip I saw only civilised dialogue and this time I did see whales. Several Minkes, mainly some distance away and somewhat prone to diving and then surfacing somewhere you are not looking. The IFAW boat crew seemed reasonably happy with the survey so far, but they were extremely unhappy at another recent rumour that whaling was due to recommence.

And so to mid-March 2013. The weather was pretty awful. A raw easterly wind, a temperature below freezing and frequent blizzards of snow, hail or freezing rain. But that was in England! Iceland was much more clement. In Reykjavík, there was evidence that there had been a snowfall not long ago, with shovelled white piles at the roadside, and white caps on the distant hills, but the temperature was several degrees warmer than in London, the skies bluer, and sometimes even sunny, and, best of all, the sea was as flat and calm as the colloquial looking-glass. Off we sailed in search of Orcas.

I have been fortunate to previously have had two pretty memorable Orca experiences. One was in Canada, on Vancouver Island, where the local marine expert has rigged up a network of underwater microphones (hydrophones) out in the bay and quite some distance apart. I sat with him on a couch in his front room as if we were listening to music on his speakers. What we heard was the music of the whales somewhere out in the bay, but 'coming our way!' Of course, not even 'Free Willy' could have leapt through the bay window, but we were able to anticipate the Orcas' route, race to the jetty, leap into a speedboat and land – or rather be thrown out – on what we had been told was a small island, but turned out to be little more than a rock! The cameraman and I were almost swept into the water by the swell as a small pod of Orcas hurtled past us like glossy torpedoes, while the sound recordist was lost in his earphones, beaming with delight as he captured enough whale songs to release an album. Cetaceans really ought to get royalties from New Age relaxation records.

My next Orca experience showed the rather more ruthless side of Killer Whales and also why they deserve that name. We were filming on the Valdes Peninsula in Patagonia, Argentina. Most of the resident wildlife is not difficult to film. Penguins either snooze by their burrows or waddle off to the beach no faster than Charlie Chaplin. Elephant seals behave every bit as badly as they look as though they should. They either lie there farting very loudly, or every now and then a couple of massively misshapen bulls will try to bite each other's heads off. 'Normal' seals were much more pleasing to look at, and to smell. Generally, though, they were not much more active, largely lounging on the sand, occasionally nuzzling their cuddly little offspring, or even nudging them to go for a paddle in the shallows. Not always a wise idea.

The action that I and the film crew witnessed was unplanned and unexpected. We were literally packing the gear into our van, when someone – I think it was me – spotted a distant fin slicing through the water, parallel to the coast. I held my breath long enough to be sure I wasn't hallucinating. Then yelled 'Orcas! Coming this way!' The cameraman and sound recordist unpacked and set up their gear quicker than the gunners at the Edinburgh Tattoo. There were four or five fins proceeding with menace, in what was surely a hunting party. The animal that was about to be hunted seemed totally oblivious. Some of the larger seals were lolloping up the shingle to safety. But one didn't. A little one, a youngster, snuggled in the tideline foam, as cosy as in a duvet. The camera lens swung left. One of the larger whales had split from the group and was surging forward as if he had slipped into a higher gear. As he headed for the shore, he sank under the water. I whispered to the camera: 'I know what might happen, and I sort of want to see it, but then again, I don't. Maybe it'll escape.' At which moment, the cameraman panned onto the baby seal at the tideline. There were three seconds of nothing. Then the Orca leapt into the picture, grabbed the seal and started to shake it violently. Spray spurted everywhere. Soon there would be blood. Killer Whale indeed.

We expected the whale to slide back with its prey, as sleek and slick as it had come. But it didn't. Then we realised that it couldn't. It was probably an inexperienced male. Young and foolish, it had grounded itself on the sandbank and now it was panicking. It had no choice but to grant me my wish. With one last flourish it slung the seal pup back on the beach, and with a final flail and an ungainly lurch, it returned to the sea, and sank out of sight before the rest of the pod could start teasing it. Sometimes you get what you wish for!

All whales are special. Each species is different. Some barely or rarely break the surface. Others wave to you with their gigantic tails or leap into the air causing a cascade and a mighty splash. Some are still hunted. None should be.

Happily, Orcas are more suited to catching food than becoming it, though they have their problems, not least from ever-increasing noise pollution in the world's oceans. Drop a hydrophone into the water and listen on earphones. The first time I did that – in the Moray Firth in Scotland – I was half-deafened by a noise that sounded like Status Quo tuning up, or a race day at Brands Hatch. It was actually the engine of a single ship over half a mile away. Imagine the confusion and chaos this must cause to the subtle and complex communications of Orcas and other cetaceans.

And so to my most recent experience of Orcas, only (as I write) a couple of weeks ago. This time our boat sailed from somewhere I can neither remember, spell nor pronounce, but it wasn't a long way from Reykjavík. I have learnt to ration my optimism when setting out whale-watching, but folks around me showed no such restraint. In fact, I have never known such confidence. I was assured that 'the sea will not be rough' – and it wasn't. I was told 'we will see Orcas' – and we did! Just don't ask me where!

I do know that for the past two years huge shoals of herring have appeared so close to land that at one period thousands of them beached on the shore, at first providing great scavenging for gulls and White-tailed Eagles, then feed for domestic animals, and finally disintegrating into a pretty disgusting eye-and-nose-sore! Orcas would of course demand fresher fish. And for two years they have been getting them as the shoals of healthy herring have increased. And so too have the number of Orcas.

Whale-watching has never been easier. We simply sailed into the shelter of the bay, cut the engine and dropped a hydrophone to provide the live soundtrack of clicks, cries, howls, barks and no doubt all sorts of sounds that are beyond our frequencies. Are they just echolocation and contact calls? Or is it a language? Is it gossip? It was tempting to believe the latter. After all, there were quite a lot of them. Counting wasn't easy as they rose and dived, whacked fish with their tails, grabbed them in their teeth and finally swallowed them before popping up again, waving their dorsal fins. When several do that at the same time it looks like the bay has been invaded by a regatta of black sails. How many did we see? I'd say at least 30.

By the end of March, the Orcas will have refuelled with herring and there will come a day when the whale-watching boat will be disappointed. We don't know where the Orcas go. Surely not to Vancouver? Patagonia maybe? Or do they swim south-east, to South Africa, Sri Lanka or Malaysia? Orcas have an almost worldwide distribution. Just like us. We can't even be certain that they will return to Iceland later in the year. We hope they will. We hope that people will be able to enjoy them. And we hope that whaling will soon become a thing of the past.

IT'S NOT ONLY THE CORMORANTS THAT CAN'T FLY.....

CHAPTER TEN

Galapagoing

I first saw the Galápagos in black and white. When I was a lad, it seems that hardly a week went by without a new Darwin drama/documentary on telly, usually called something like 'The Voyage of the Beagle', or the 'Origin of Species', or maybe just 'Darwin'. They weren't what we now know as natural history programmes. They featured actors playing historical characters, wearing period costume and jotting in their journals while tiptoeing through the iguanas. Sort of *Life on Earth* meets *Downton Abbey*.

Thinking back, it is pretty amazing that these shows were filmed on the actual location. It wasn't the Scillies pretending to be the Galápagos. Nor was it possible to hire the wildlife from Animal Actors. A few sealions could have been procured from an aquarium, but I know of no pet shop in the world where you can buy a full set of Darwin's finches. These riveting programmes really were created out in the Pacific Ocean, 600 miles west of Ecuador and – especially with the advent of colour television – we could all see that the Galápagos archipelago was a truly extraordinary place. Not only were many of the species endemic and extremely rare, but they lived in a bizarre volcanic landscape. The Galápagos seemed to me to combine qualities that were intrinsic to maximum wildlife enjoyment: a context of characterful or impressive scenery, an abundance of creatures and a scarcity – or complete absence – of human beings.

The only people we saw on these TV 'reconstructions' were the actors portraying Mr Darwin, a few of his companions, and perhaps a misguided mariner who caught and cooked a Giant Tortoise for supper, and thus provoked a lecture from Mr D on the sacrosanct uniqueness of the island creatures. This scene would be followed by Darwin wandering pensively alone until his attention is drawn to a small flock of what look like female House Sparrows. He notes and sketches their variable beak sizes and emits a quizzical 'hmmmm'. An historic revelation is nigh! At which point, the soundtrack soars, the credits roll and we have to wait till next week's episode. For many years this was my vision of the Galápagos: a remote haven where wildlife could bask, swim, fly and multiply, and where a man could be alone.

That vision was still in my head, when in 1986 I accepted an invitation from a travel company to join a party of a dozen journalists on a small boat taking a short cruise round the Galápagos. Such a thing is known as 'a press trip'. Not to be confused with a press-gang, though in this case there were similarities. I had envisaged flying on a petite 'Island Hopper' plane, like the ones that have ferried me round Shetland or The Seychelles. Instead, we were herded onto a Boeing 727. Three hours later we were herded off at Baltra airstrip, not certain if we were guests or prisoners. We were lined up by a rather stern young woman in a bright yellow jacket and matching skirt who informed – or warned – us that once we had cleared passport control, we would be shouted at by men with megaphones. We must listen for our name and follow our man.

There were several men, but no megaphones. Instead, they all shouted names in the general direction of the tourists who had arrived on the 727. They were typical tourists. No recognisable scientists, no TV presenters – not even David Attenborough before he was a 'Sir' – and no actor in a frock coat giving us his Darwin. I was disappointed. I also had a feeling of foreboding. Over the following three and a half days, the disappointment increased and the feeling was fulfilled.

It started well enough. Within not much more than an hour, we were being encouraged to hop carefully from a gangplank onto the slippery rocks of a seabird island. It was not dissimilar from landing on Inner Farne, except that instead of Grey Seals bobbing around our boat, there were sealions. Black-headed Gulls were replaced by Lava Gulls (very rare) and Swallow-tailed Gulls (very beautiful), while lowering above us were massive frigatebirds, hooked beaks and jet black, like a cross between a vulture and a pterodactyl. Most of us stopped to gaze upwards and around, but our reverie was shattered by a voice that sounded like a female sergeant major. 'Attention! Keep up please. Keep up!' The lady in yellow who had met as at the airport was not only the tour company 'rep', she was also the tour guide. She shouted out a few names: 'Blue-footed Booby, Marine Iguana'. I ungraciously whispered to one of the journalists: 'A yellow twin set? Is she an air hostess or a chalet girl from *Hi-de-Hi!*?' 'She's certainly not a wildlife expert,' muttered another journalist, who had failed to extract any information from her that was any different from what was in the official guidebook. I sighed, found myself a comfy rock and crouched down to enjoy the company of a pair of Blue-footed Boobies, in the hope of photographing their high-stepping mating dance – surely the inspiration for John Cleese's silly walk – but no sooner had I adjusted my f-stop than I was summoned by 'teacher'. 'Hurry up, please! Get a move on!'

This I was not happy about. I was even less thrilled when the next day we sailed six hours to the Darwin Centre, where we barely had time to make Lonesome George feel less lonesome (sadly, of course, he is now an ex-Giant Tortoise) before we set sail for another four-hour journey to a beach where the number of iguanas was almost matched by the number of tourists, each group being herded and chivvied by lemon-suited guides. It wasn't crowded, but it certainly wasn't secluded.

Frankly, I was beginning to feel hassled. So were most of my fellow journalists. Murmurings of discontent evolved into words of protest when the next day's schedule was announced. 'We will sail north to the island of the Waved Albatross.'

'How long will it take?' came a nervous enquiry. Followed by: 'Will the sea be calmer? 'Cos it was very choppy today.' Someone else did the maths: 'So, that's 10, plus eight or nine... That's nearly 20 hours on the boat!' And about four hours on land.' It is not often that you get a mutiny on a wildlife cruise! We didn't get quite as far as ousting the skipper – or the guide – but we did refuse to sail further north.

The yellow lady protested: 'But you will miss the albatross.' Frankly, the only one who seemed to care was me, but I didn't care enough to endure another day retching or befuddled by Stugeron. Nevertheless, sensing that I may well have been the only 'heavy' birder in our group, she targeted me with temptation. 'Don't you want to see the Flightless Cormorant?'

'It's a cormorant and it can't fly. I'll imagine it.'

'OK, but what about Darwin's finches?' she pleaded, as if she was offering me the crown jewels. This, despite the fact that when I had requested elucidation on finch identification she had been totally bamboozled. Mind you, no great shame there. There are three ground finches: the Large Ground Finch, the Medium Ground Finch and – you guessed it – the Small Ground Finch. Straightforward enough you might think, except that according to the book: 'They are all identical in plumage, and each species varies in size'! I did enquire about the whereabouts of the Vegetarian Finch, the Vampire Finch and the Woodpecker Finch, but got no assurance that we'd see any of them if we sailed for nine days, let alone nine hours. I suspect she had never heard of them, but I am not making them up, honest.

And so the mutineers took over. We spent a leisurely day taking photos (there is nowhere easier in the world) and a delightful and genuinely calm evening moored in a blue bay with orange rocks, where journalists swam with the sealions, and I strolled along the strand listening to mockingbirds. For half an hour it was as if I had the Galápagos to myself.

That was in 1986; what is it like now? More hotels, more boats, more tourists? No doubt. More wildlife? Happily – thanks to the Ecuadorian Government and international designation – probably much the same. Problems with goats, cats, rats, etc? Same as it ever was. If you want to find out more there is masses on the internet. I am not the one to ask. I have never been back. Will I? Would I? Well, maybe if… Frock coats do rather suit me. And I've got the beard for it. My Darwin could have a new catchphrase: 'I want to be alone.' Or how about: 'We're gonna need a bigger boat!'

BIRDS ON THE BOX

CHAPTER ELEVEN

They Couldn't Do That Now!

Picture this. A small suburban lawn. In the middle, there is half a tree. The trunk is riddled with cracks, crevices and holes. One hole is particularly conspicuous. It is only a couple of metres from the ground. If it were higher up, in a bigger tree – with branches – it would surely belong to a woodpecker. But not… At which point, we hear the impatient *tchek tchek* of an adult Great Spotted returning to the hole, and the frantic squeaking of fledglings inside it. What a set up! A photo or film opportunity if ever there was one. Only one problem, how do you get pictures from inside half a tree?

First, find a man with a steady hand and an electric saw. Slice a sliver of wood off the back of the tree trunk, thus exposing the nest chamber, and the fledglings. They can't fly yet, but they could fall out, so be ready to catch them. Now replace the sliver with a pane of clear glass. Next, rapidly erect a canvas hide at the back of the tree, so that a cameraman's lens will be able to observe and film whatever is happening through the glass. Now retire to the nearby kitchen and enjoy a nice cup of tea with Enid, the lady of the house. The kitchen window overlooks the lawn. Very soon Enid will call out: 'They're back. Woody and Winnie! Wood! Win! Yoohoo!' You might have thought that Enid's calling and waving might have unsettled the birds even more than the electric saw and the hide installation, but the fact is that once the eggs have hatched most parent birds exhibit extraordinary devotion, tolerance and bravery in caring for their offspring.

Nothing was going to deter Woody and Winnie from rearing their young, certainly not the sporadic rustling of a cameraman crawling into the hide, and setting up his enormous tripod, and focusing his lens through the little window and revealing three chicks that were so young they were not yet either 'great' nor 'spotted', or indeed even slightly feathered.

As days passed, they grew bigger, and turned from baby grey to adult black and white. The only colour in the scene was the yellowy glow of the cameraman's lamp through the glass, towards which the chicks duly directed their rumps. Birds that nest in the dark defecate towards the nearest light. This is usually the nest hole, but not in this case. The cameraman had to do a lot of wiping. Nevertheless, he got some fabulous footage, arguably verging on the historic.

But when – you may be wondering – was this? Was it the work of one of the great pioneers of wildlife filming? Cherry Kearton perhaps? Or maybe Heinz Sielmann – he definitely did woodpeckers. Probably using much the same techniques as we did. We? I mean, of course, the BBC Natural History Unit. Woody and Winnie Woodpecker featured on a series called *Bird in the Nest*, which was broadcast live in 1994, and again in 1995. The presenters, claustrophobically crammed into a small van – the 'Birdmobile' – were Peter Holden (from the RSPB) and myself. Among the ace cameramen/naturalists were Charlie Hamilton James and Simon King.

Contemporary equivalents of *Bird in the Nest* can now use astonishingly minuscule cameras that can easily be inserted into holes and burrows – or even the wildlife itself! – but they are also bound by strict rules and guidelines about what can and can't be done, not to mention the ubiquitous 'health and safety', for people and for wildlife.

Not so back in the mists of the mid-1990s. Nest set-ups were achieved by whatever disconcertingly ingenious means was necessary. Filming Kestrels involved erecting a precarious scaffolding tower. Jackdaws were persuaded to nest inside an empty oil barrel, suspended at a considerable height. A Great Tit family in a nestbox were removed from a tree, transported across the garden and relocated on the side of a shed, which had been erected specially for the purpose. It was big enough to accommodate a cameraman, or indeed a presenter. Peter nearly missed the start of one show because he'd got locked in with the tits!

First-time-ever live pictures of newly hatched Kingfishers were achieved by a similar technique to the woodpeckers, except that instead of sawing a slice off a tree trunk, Simon dug a pit in a riverbank. Viewers were enthralled when the tiny, blind and naked chicks made their historic debut, but by morning they were incensed and distressed by the news that, later that evening, the parent birds had been spooked by a band of boisterous locals on a Treasure Hunt, which had involved a deal of shouting and splashing up and down the river. The chicks were left unbrooded for maybe 'only' an hour or two, but they chilled, and by morning they were dead.

Misfortune also befell Woody Woodpecker. We had often worried that his regular flight path involved flying across the nearby busy road. His typical woodpecker undulating flight style meant that at the highest point he was safe, but at the lowest he was definitely potential roadkill. One day, the inevitable happened. Or so we believe. A lorry driver came to see us and – with a sad and guilty face – confessed: 'I think I've just run over your woodpecker.' We tried to console him by saying that it might not be Woody, but we all knew it was. Especially because he was never seen again. Peter and I rhetorically asked the viewing public whether or not we should provide extra food for Winnie and her rapidly growing brood. These days the question would have been literal: 'Shall we save them by putting out mealworms? Or shall we let nature take its course? You decide!' A bowl of mealworms was placed just below the nest hole, so near that all Winnie had to do was lean out and grab a beakful. As the chicks grew, they learnt to do the same. They survived, and the viewers rejoiced just as they had wept for the Kingfishers.

Bird in the Nest was dubbed 'an avian soap opera', and great accolades were bestowed on the BBC by the national press. But I can't help thinking: it's a good job they didn't find out how some of it was done!

CHAPTER TWELVE

Anthropothingy

Anthropomorphism. Some people can't say it. Some people can't spell it. And some people can't stand it. Especially when the presenter of a wildlife film or TV programme starts giving creatures human voices, thoughts, opinions and emotions. I first realised that it was frowned upon by 'serious naturalists' when I attended my very first Wildscreen many, many years ago. The commentary on one of the films involved a slight tinge of the 'Johnny Morrises', which the majority of the audience received with what I could only describe as a collective sneer! I did not condone this derision, as must have become clear from the fact that on my own programmes I have rarely resisted dabbling in the anthropomorphic myself.

Why? Because it is challenging – one might call it acting – fun to do and, hopefully, to hear. It may simply be diverting, rather like that fanciful question: 'If you were an animal what would you be?', or it might help to illuminate animal behaviour. Watching Black Grouse on a lek, I found myself giving them dialogue like a couple of hard lads taunting each other, hoping to impress the girls by their bravura, while never actually coming to blows. 'Right mate, you want some do you?' 'Think you're well 'ard don't ya?' Anticipating any accusations of trivialisation, my producer – bless him – provided the perfect defence, by confirming that such a 'ritual' stand-off was exactly what the cock grouse were doing. What's more, to add to the anthropomorphic accuracy, a couple of disinterested hens stood at the side muttering: 'I don't fancy yours much!' The only complaint I got was that, though the lek was in Scotland, I gave them all cockney accents.

The lads and lasses at a club could validly be compared with what happens at a Black Grouse lek – and vice versa – and there are a few more basic urges, needs and experiences shared by both humans and animals, such as hunger, starvation, pain, eating, sleeping and sex. However, it is attributing subtler human emotions and characteristics to wildlife, such as jealousy, nostalgia, depression or happiness, that is usually deemed 'dangerous' by scientists. 'Dangerous' because it may well be wrong or at least not provably right!

Where anthropomorphism really comes into its own is when the words and thoughts are being put into the heads, mouths or beaks of animals in fiction. How authentic they are is the author's decision. It is accepted that *The Wind in the Willows* celebrates Kenneth Grahame's love of the river and its wildlife, but don't look to it as a compendium of accurate animal behaviour! Badger lives alone, real badgers live in groups. Moley potters around in daylight, real moles are strictly nocturnal. Ratty is clearly a Water Vole, and toads are slow, secretive and almost silent. No wild creature rows a boat or drives a vintage car. None of them talk. Grahame's creativity and characterisation were inspired by observing his fellow men, not just wildlife.

The fact is that animals and humans are a formidable combination. From fables to films, they provide the looks, we provide the words! A few of us provide the talent to draw, paint or animate. Consider the number of cartoons – full length and 'shorts' – that feature eloquent animals as the principal characters who, by the way, effortlessly upstage the humans. Is there anybody who considers Mowgli to be the star of *The Jungle Book*? And can you imagine anyone else – animal or human – performing 'Bear

Necessities' better than Baloo? I happily admit that few moments delight me more than when a cartoon animal character bursts into song. Even better if they dance, and better still if they are joined by a chorus. Surely we would all echo Dr Dolittle: 'If I could talk to the animals… And they could talk to me.' Except I would change that to 'sing'!

I believe that many of the greatest works of art – or entertainment? – are what I would call 'animated anthropomorphism' (though I might not be able to say it after a couple of double Scotches). Wild creatures may be fascinating and photogenic but, let's face it, they are not particularly creative! They may make us laugh and go 'aw', but they don't intend to, and – as far as we know – they have no imagination. I'll own up, I prefer my rabbits to tap dance and play the banjo (though I reckon that Bugs Bunny is really a hare). I am happy when my bluebirds harmonise on 'Zip-a-dee-doo-dah', and when I saw my first real Roadrunner and it didn't go 'beep beep', I felt like complaining to the Creator.

However, let me be clear, I am not giving up wildlife and, if any creature is feeling under-appreciated, let me remind us all of one thing. Without animals, there would be nothing to be anthropomorphic about.

Indeed, there would probably have been no human creativity at all. No music, dance, art, literature, drama, animation, nor even any high definition or 3D! In all things, man was surely inspired by nature. There was nothing else.

CHAPTER THIRTEEN

I Know That Tune

Everyone knows that the playback speed of recorded sound makes a difference to its pitch. Fast equals high pitch, slow equals low. Mind you, it is not so easy to demonstrate on a CD or an MP3. It was much easier with vinyl. Play a 45rpm at 78rpm and it sounded like Pinky or Perky. Play it at 33rpm and it became the voice of God. Is it just a bit of aural fun, or does it have deeper implications?

A couple of years ago, I visited the studios of Andy Sheppard, a world-class jazz saxophone player, who is also renowned for incorporating natural sounds in his music. He told me that his initial fascination was born of irritation. He had been playing at a club in Germany, and such is a jazz musician's nocturnal schedule, he finally got to bed as dawn was breaking and the birds were beginning to sing, including a Blackbird that was so loud it kept Andy awake. However, such was the loveliness of the song, that instead of cursing it he recorded it, which was such a relaxing process that he nodded off.

On waking many hours later, Andy reached out of bed and clicked on his tape recorder. What he heard literally scared him to death. Instead of the mellifluous melody of a Blackbird, his earphones were filled with a low rumbling roar, 'like dinosaurs in the movies!' He didn't recall recording the soundtrack of *Jurassic Park*. What he'd done was accidentally playback the Blackbird at a much slower speed, which transformed it into a *T. rex*! He increased the speed – not a huge amount – and the dinosaur turned into a whale. A little faster again, and it became a dolphin, until – back at normal speed – it reverted into a Blackbird.

As Andy altered the speed and pitch, I listened with my eyes closed. Every sound was exactly as he'd said. Finally, I asked him to speed up a recording of actual whale song. It sounded like a Blackbird!

I am not going to attempt any scientific theory or analysis of natural sounds, what they mean, why they are as they are, and whether they are random or rational. For that I refer you to *The Great Animal Orchestra* by Bernie Krause. Krause often refers to the work of British sound recordist Chris Watson. I have worked with Chris for many years and have heard many extraordinary things 'before my very ears'.

For example, the day he turned me into a Wren. I have the figure – small and dumpy – but what about the voice? A Wren's song is not long, but it is very fast, packed with notes and incorporating a characteristic trill. There was no way I could whistle or sing it. Then Chris slowed it down. What amazed me was that many of the notes swooped, whooped and bent, more in the manner of a dolphin or a gibbon. What to our ears was a single note was in fact several. The trill – unsingable for a human – became a slow-paced *tuk katuk katuk katuk*, which I reckoned I could get my larynx around. I prepared to sing like a Wren.

For half an hour, I listened, learnt and practised. Then Chris recorded me doing my slow-speed Wren impression. He played it back and I sounded somewhere between a scat singer and a human beatbox. Then, the moment of truth, to play it back at Wren speed. It became faster and higher-pitched. You only have my word for it, but it wasn't half bad! I was so confident, I perched on a gatepost, miming to the recording. At least one Wren sang back, but we drew the line at mating.

CHAPTER FOURTEEN

Don't Look Now

So there I was, nibbling my croissant, when in storms this enormous American woman. 'Mr Oddie, Mr Oddie! The swans are attacking the Canada Geese!' I was of course taken aback. It's not often I meet anyone feeling sorry for Canada Geese. Added to which, what did she want me to do? 'You've got to stop them!' Did she seriously want me to go swan wrestling? Wasn't she aware of the folklore? 'They can break a man's arm, you know.' Not that I have ever heard of one that did. Nevertheless, I bet they could make mincemeat of a Canada Goose! But I didn't tell the lady that. Instead, I resorted to euphemisms.

I explained that 'Mute Swans are very "territorial", as well as "protective" and "faithful". All admirable qualities, I think you'll agree.' She didn't. 'But they are killing the Canadas!' I countered that the appropriate word was not 'killing', but 'deterring'. 'They are just trying to make the geese go away by…'

'By killing them!' I was not winning. Clearly I would have to resort to the adjective that will brook no censure. 'Madam, what those swans are doing is entirely *natural*.' For a moment, she was silent, no doubt unable to resist her reverence for the blessed 'N' word. Her next pronouncement was not what I expected:

'Well, it may be natural, but I don't want to see it!'

'So don't look!' I suggested.

As she stomped out of the café, the thought struck me: I bet she doesn't watch many wildlife programmes!

Most of what wildlife programme-makers call 'animal behaviour', in a human context would be '*mis*behaviour'. Such content in a drama is preceded by a warning. (Or is it a promise?) 'The following programme contains scenes of a sexual nature, violence and bad language from the start.' The same ingredients in a wildlife programme are just what the audience expects to see, and certainly what the production team loves working on. A wildlife cameraman would kill for a kill, an editor loves cutting a chase sequence, the sound recordist revels in capturing every roar, and the director can't wait to add some Barry White to the soundtrack for mating hippos. The final cut – the broadcast version – is approved by the producer, or someone even higher, so if viewers disapprove, that's who to complain to!

But you can't edit a live programme. I recall several times on *Springwatch* when there was adverse audience reaction. The pair of House Sparrows that was guilty of exhibitionist coupling; the Blue Tit that brandished the corpse of its dead chick, before ejecting it from the nestbox; and the famished adolescent Barn Owl who staved off starvation only by gobbling up his baby brother. Yes, there were some complaints, but it was nobody's 'fault'. The show was live. We didn't know what was coming. But if we *had* known, would they have been shown? I suspect they would. But there was one that definitely wasn't.

It was on *Wild in Your Garden*, live, from Bristol. There were cameramen everywhere, secreted behind bins and fences, and in sheds and hides. One of them was Dave (not his real name). It was only when I saw him in the supper queue that I remembered he'd been staked out near a birdtable all day. Obviously, the control room had completely forgotten he was there! 'It's a pity,' he grumbled, 'because I got some great Sparrowhawk action.' He invited me to watch a playback. It started with the hawk perched on a fence. Then it took off. Dave followed its every twist and swoop until it grabbed a Starling in midair. It then flapped off to a fence post and – holding the Starling down with one foot – and gripping its frantically flailing wings with the other, it proceeded to rip at the Starling's chest with a bill like a meat hook, and began eating it alive!

What did that lady say? 'It might be natural, but I don't want to see it.' Nobody did.

INFINITE VARIETY

CHAPTER FIFTEEN

Out for the Count

'The hobby that dare not speak its name.'

Well, that's how it used to be. By the time I was seven, I knew I was a birdwatcher. I didn't admit it till I was 10. The response of my peers was puzzlement, derision and mockery. 'Birdwatching? How soppy! Why can't you go scrumping apples and pulling girls' pigtails like us normal schoolboys?' Adults were just as indifferent, but men could rarely resist coming out with that most tedious of 'bird jokes': 'Oh yes. I'm a bit of a birdwatcher too! Two-legged kind, eh? Eh?' I endured that several hundred times, before I finally discovered the deserved riposte, which is: 'All birds have two legs. Unless they've been in an accident and lost one, in which case it's very distasteful to laugh at crippled creatures.'

The 'I'm a bit of a birdwatcher too' quip is still going pathetically strong, but the joyous truth these days is that many people *are*. 'What, the feathered kind?' Yes. Indeed, things have changed so much since I were a lad that, instead of being an esoteric minority activity, birdwatching is arguably the fastest-growing leisure pursuit in the world. Well, that's what I read in the business section of an American Airlines in-flight magazine about five years ago. It has probably been overtaken since by ballroom dancing and tweeting, but nevertheless it is undeniable that birdwatchers are no longer alone. We are out – outdoors that is – and proud. No longer are little lads and lasses teased about their hobby. Parents encourage them. A lot of them join in.

However, birdwatchers are not all unified as one harmonious band, but that is no bad thing. One of the delights and allures of birds is that they can be enjoyed in so many ways. Some people draw and paint them. Others photograph and film them. Some record their songs. Others 'twitch'. By the way – press and media please note – 'twitcher' is not simply a synonym for 'birdwatcher'. In the same way that a sprinter is an athlete, but an athlete is not necessarily a sprinter, a twitcher is a birdwatcher, but a birdwatcher is not necessarily a twitcher. Twitching is the often rather frantic pursuit of rare birds. We've all been on an occasional twitch, but a serious, knowledgeable birdwatcher who is not obsessed with his or her 'list' would prefer to be called 'a birder'. I am a birder.

I am not a 'bird spotter', an expression that belongs in pre-War Boy Scout manuals and I-Spy books. Nor would I claim to be an 'ornithologist', a title which implies scientific knowledge, a capacity for protracted study, an understanding of graphs, figures and statistics, and possibly a doctorate. Finally, at the opposite – but not bottom – end of the birdwatcher's league are people who put out bird food in their gardens, may not even be able to identify all the species, but simply enjoy having them there. Let's just call them bird lovers. Actually, we are all bird lovers aren't we? Birdwatchers, twitchers, ornithologists, birders, even bird spotters (if they are not extinct), we need you all.

At various times of the year, the RSPB, the BTO, the Wildlife Trusts and others organise surveys and events which you can take part in whatever you call yourself! Visit the websites for full details, including help with identification.

Facts, figures and fun.

This piece helped to publicise the RSPB's annual Big Garden Birdwatch.

CHAPTER SIXTEEN

Blowing in the Wind

'Twitchers flock to see rare bird!' So says the headline above a photo of a couple of hundred bearded blokes with binoculars. But what actually constitutes rarity? Well, in this context we are not talking 'endangered' (apart from being trampled by twitchers) nor does it refer to a scarce breeding species. The rare bird that hits the headlines and attracts the crowds is usually on its own, and probably confused. A rare bird is a lost bird. Alternative names confirm just that: an 'accidental', a 'vagrant', a 'drift migrant'. That one sounds almost romantic, but it isn't. A bird's migratory journey is a risky enough business, without wandering off-course and ending up where it didn't mean to go. Not surprisingly, it mainly happens to long-distance migrants. The ones that boggle us by the distances they fly and the routes they take, across oceans, deserts, mountains, forests and cities. Why do they get lost? Sometimes it is a malfunction of their navigation system. A dodgy satnav. Mainly, it is the effect of extreme weather. Sometimes extremely *good* weather.

In spring, a migrant flying north may get literally carried away by the ease and joy of soaring in clear blue skies on a warm southerly breeze and unwittingly overshoot its intended destination. A warbler that meant to nest in France may end up in Shetland. It won't find a mate. It won't find its way back either. It is lost, it is rare, it may well get twitched.

In autumn, it is most likely to be strong winds that are the problem. Migrants get blown off-course. Powerful air streams from the east, from Europe and beyond may transport birds from as far away as central Russia or Siberia. Birdwatchers call these birds 'sibes'.

But most of our wind comes from the west, sometimes originating in America and crossing the Atlantic, heading for the UK. On 6 September, I was heading for the Isles of Scilly, off Cornwall. At the same time, racing north-east off the coast of North Carolina, was Hurricane Irene, or was it Kate? Or Katia? I don't remember her name, but she was a fast lady. In not much over a day she swept across 'the pond', expending so much energy that she was demoted to a 'tropical storm', and then veered north to give the Hebrides and western Scotland a good lashing and a gigantic cold shower.

In Scilly, we only felt Katia's tail-end, but that was frisky enough to make my walking against the wind far more authentic than anything Marcel Marceau ever achieved. I headed for calm in the lee of a hill, suspecting that any sensible birds would do the same. I felt barely a breeze as I leant on a five-bar gate and scanned the field in front of me. It was a grassy field. Not long grass, but not very short either. Bisecting it was what looked like a wide bare earth path, which I later discovered was the aftermath of recent pipe burying. The scar was sandy, with a few pebbles and a shallow puddle. 'Wheatear habitat,' I predicted as I scanned. But there were no wheatears. 'OK, then, wagtails.' Sure enough, a White Wagtail trotted out and duly wagged its tail. 'And, what is that?' A little bundle of pearly-edged feathers. Asleep? Exhausted, more like. A Buff-breasted Sandpiper. Just flown in – or blown in – from America. Or rather from Arctic Canada. That is where Buff-breasts breed. Most of them migrate south to Argentina down the

Great Plains flyway, but most years a few reach Britain and Ireland, and Scilly.

As this fragile little traveller sighted the islands it must have sought out somewhere it felt comfortable, somewhere that reminded it of home, a tiny patch of surrogate tundra, a scratchy path scoured by the local pipe-layer, and a puddle just like the ones left by melting snow. OK, not exactly the Yukon, nevertheless comfortingly familiar. If only it had carried on for another mile or two to the island of St Mary's and checked out the airport, it would have found lashings of close-mown grass and another seven Buff-breasted Sandpipers! Next day, it joined them. Eight Buff-breasts together! That's not a vagrant, it's a flock!

Which makes me wonder, do they really all get blown over, and then find each other by luck? Or do they mean to come the transatlantic route, having prearranged a rendezvous? Great Plains? Boring! See you in Scilly.

CHAPTER SEVENTEEN

Puffin Billy

Back in the 1990s, an American friend and his wife came over from New York to spend Christmas in Britain. In early December, he rang me and announced: 'We have decided to give ourselves an early Christmas present. We want to see a Puffin. So, your present to us is to tell us where to go, as it were (some Americans *do* do irony). Can you do that?' 'No,' I replied firmly. He was understandably taken aback. Was divulging the whereabouts of Puffins a threat to national security perhaps? He confessed he'd been doing some research. 'What about Lundy island? *Lundi* means Puffin doesn't it?'

'Yes, but you won't see any there.'

'Shetland?'

'No.'

'The Fern Islands?'

'Farne Islands. Best place in Britain to get really close to seabirds.'

'So can't we go there?'

'You can, but you won't see any Puffins. Not at Christmas. But if you can stick around till Easter...' At which point, he figured it out. 'Ah, so Puffins breed in Britain, but they don't winter there.'

'You got it.'

'So where do they winter?'

'I don't know. Nobody knows. Presumably out in the ocean. Probably in the Atlantic.'

'Only probably?'

'We are not even sure about that. The truth is we simply have no idea where Puffins – and other seabirds – go in winter. Not that we haven't been trying to find out.'

Flashback to 1957. It is nearly Easter, and the seabirds are returning to the Farne Islands. Guillemots are gathering, lining up on the ledges like fluffy toys on rocky shelves. Razorbills cluster in crannies where they think they can't be seen. There are Puffins everywhere. Flying on wings so rapid that they blur, and zipping past me like miniature missiles, before crash-landing with less control than a manic schoolboy tumbling into the long jump pit. Like, for example, myself, in my mid-teens, 'intermediate' school athletics champion and temporary member of a Puffin-ringing expedition from Monks' House Bird Observatory on the Northumberland coast. In charge was a man with a figure like Mr Pickwick and a chuckle like Tommy Cooper, one of the greatest and most charming naturalists, artists and mentors a budding birder could wish to have. Dr Eric Ennion. They don't make 'em like that any more.

Dr E. led us nimbly over the spongy turf of Inner Farne, through what one might have assumed was an extensive rabbit warren, but was actually the Puffins' Underground City, with each subterranean apartment as snug and safe as a fall–out shelter, unless the booted foot – of say, a schoolboy like me – accidentally crashed through its grassy roof. What cruel irony if a bird that had survived six or seven months being buffeted on stormy oceans had found his way back to the very same burrow he had occupied last summer, and was now sitting there waiting for his faithful lady love to join him, paternal instincts trembling and hormones rising, when suddenly 'splat' – he is flattened by a boot from above, as unceremoniously as by Monty Python's giant foot.

Fortunately, the first hole was unoccupied, but my forfeit was that Dr Ennion appointed me as principal Puffin catcher. I was issued with a large pair of thick leather motorcycling gauntlets, which suggested that Puffins might not be as sweet and docile as they looked. Especially if they were relishing the arrival of a sexy female Puffin, but instead were molested by the leather-clad fingers of a nervous schoolboy. Believe me, it is terrifying sticking your hand into a pitch dark tunnel, not knowing whether or not there is something alive in there. It is scary if there is nothing, and even scarier if there is. You will soon know if the burrow is occupied, because the Puffin will try to pull you down there with it!

You now have two choices: you can let go, and apologise to both the Puffin and to your now derisive fellow Puffineers, or you can grab the bird with one hand and try to wrestle it to the surface, where you can use your other hand to grab its beak, which could surely snip off a finger as neatly as a pair of secateurs severing a twig. Thus deprived of its most lethal weapon, the bird appears to accept its fate and seems calm, content and even curious as you flick the gauntlet from your free hand so you can rummage in your shoulder bag for your notebook, ringing pliers and, of course, rings. You even risk relaxing your grip on the rainbow-coloured finger lopper. Such a pretty beak! And so ingeniously capable of holding more than a dozen slippery sandeels. Surely this Puffin will not peck me. Indeed it won't. But it will lacerate your wrists and forearms with its needle-sharp little claws.

It was easy to tell which of us had been on Puffin Posse duty. We looked like a bunch of extras from *True Blood*. Nevertheless, we bore our scars with pride. After all, we were contributing to science. Each bird was fitted with a lightweight metal ring inscribed with a number and the request 'Inform British Museum'. This was not so that the curators could come and recover their lost exhibit; it was an instruction to whoever found what was left of it. Most so-called 'recoveries' came from decidedly ex-Puffins: mainly tideline corpses, murdered by a Great Skua or a Great Black-backed Gull, choked by an oil spill, poisoned by pollution, strangled by a fishing net, or even – just a few – finally succumbing to old age.

Puffins can live for 35 years or more. That was something we had learnt from ringing, which also taught us that Puffins are faithful both to nest site – the very same burrow if it is available – and also to their mate. We have also compiled all kinds of data about what is often nominated as Britain's favourite bird, and which is undeniably fascinating and photogenic, and seems as relentlessly addicted to funny walks and pratfalls as Charlie Chaplin. But Puffins also have a dark side. Literally. They spend an awful lot of time underground, but – thanks to such up-to-date technology as endoscope cameras and minuscule microphones –we know a lot about their private life. However, until recently, there was still one thing we didn't know – where do they go in winter?

Fast-forward to 2013. I am on the Island of Skomer, off Pembrokeshire, South Wales. It is Easter, early spring. It's Puffin Time. Again I am in the presence of authority, but it is not a warden and his team of leaden-footed schoolboys, this time it is a revered scientist and his young French female assistant, whose footwork is as delicate as a ballerina's. A change for the better. There is another change, today's Puffins do not suffer the indignity of being dragged out of their burrows by hand. Instead, they are yanked out by foot, with a sort of miniature walking stick. Or they catch themselves in a small almost invisible 'mist net', from which they are disentangled by the scientist, or his lovely assistant, so deftly and rapidly that they barely have time to wonder what's happening.

Then comes the biggest change of all. The temporarily captive bird is not merely fitted with a ring on its leg. In addition, attached to the ring is what looks like a miniature radio. I assumed that this was some kind of tiny transmitter which would send signals back from wherever the Puffin ventured, but I was told that it was more like a midget computer. 'A sort of Puffin satnav?' I suggested.

'No,' said the scientist. 'A satnav tells you how to get somewhere.'

'Allegedly,' I interjected.

'While this device tells you where you have been', continued the scientist.

'Or rather where the Puffin's been,' I commented, sensing an impending historic revelation. 'So, does this mean we now know where Puffins go in winter?' In my head, I heard accelerating chimes building the tension. 'So what's the answer? Is it a yes or a no?'

'Well, Bill, it is… a… yes!' Blimey, cue audience applause! But hang on, this is a scientist talking. Scientists never completely commit. I bet there'll be an 'on the other hand', or a 'mind you', or – at the very least – a 'but'. Indeed there was a 'but', but it was a pretty incredible 'but'.

'We now know some of the places they go in winter, *but* they don't all go to the same place. For example, we have data that proves that some Skomer Puffins winter mainly off Portugal, while others winter off Newfoundland.'

'Birds from the same colony winter in different places?'

'Exactly. In fact, we had two birds in adjacent burrows. One went to Portugal, while next door went to Newfoundland.'

Now, I was full of questions.

'Do they go to the same place every year? Do male and female go to the same place? Will they meet up with Farne Island Puffins? And come to think of it… why? Why?!'

The answer was inevitable: 'We don't know. Yet!'

That evening I emailed my friend in New York. 'Hi Harvey, remember you wanted to see Puffins for Christmas and you were going to come over to Britain? Well, if only we'd known: you only had to go to Newfoundland!' He responded immediately: 'Christmas in Newfoundland! You gotta be kidding me! No way!'

'OK,' I replied, 'how about Portugal?' I attached a link to 'Winter Breaks in the Algarve'. The brochure had pictures of girls in bikinis, bottles of Mateus rosé, sardines on toast, and even a pair of White Storks, but no Puffins. Somebody should tell the tourist board. Harvey still hasn't seen one.

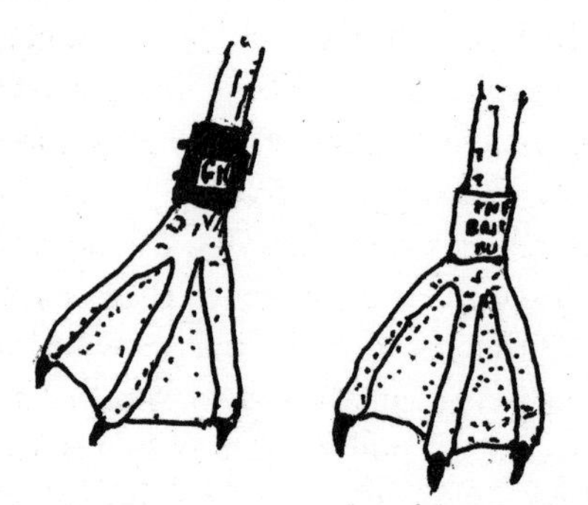

WHY NOT
MIX & MATCH?

CHAPTER EIGHTEEN

Flying Tonight

You know when tourists float across the African savanna looking down on the wildlife from the basket of a hot-air balloon? It has always bothered me. Not that I have had personal experience. The only balloon flight I have ever taken was over Loch Lomond in Scotland but, as it turned out, I was pretty bothered by that one too. It was both scary and farcical. The wind was very light. We took off vertically upwards, drifted slowly out over the water, and then began plummeting vertically downwards, seemingly destined to pitch in the drink. There were quips about being eaten by salmon and maybe even beavers (surely they have been reintroduced up there?). Such jests were lost on me. I can't swim.

Fortunately, our pilot (they do it by willpower!) managed to steer us back over land at the last minute, where we plonked unceremoniously down in a field so glutinously boggy that a local farmer had to be summoned to pull us out with his tractor. However, no sooner had he attached a rope from his vehicle to the basket than the balloon began to rise again, so abruptly that it lifted the tractor clean off the ground, with the farmer still sitting in the driving seat! It was a scene worthy of a Buster Keaton movie (or dare I say *The Goodies*?) but became less laughable when we realised that the dangling tractor, the farmer, the balloon, the basket and us were now being swept directly towards an electricity pylon and power lines. Happily, the breeze dropped just enough for the ever-increasing crowd from the village to stop laughing, slurp across the bog, grab hold of the tractor and pull us all back to safety.

But what has this to do with wildlife, apart from the fact that we were filming for the Natural History Unit (so much for health and safety!)? The point is this: while we were noiselessly floating over the farm fields, the animals below seemed to barely notice us. However, when the balloon went for a 'burn', horses, cows and sheep scattered and panicked as if they feared imminent attack from a giant winged predator. A balloon burn emits a roar that a *T. rex* would be proud of. Of course, the wildlife of the Serengeti and the Masai Mara is well used to roaring sounds, so maybe the balloon burns don't bother them, especially since hopefully Kenyan balloons are less prone to plummeting than our Scottish version and are therefore considered less of a threat. Nevertheless, I am still sufficiently bothered by the African balloon safari ride to have never taken one.

I have, however, had a Kenyan aerial experience that was surely much more memorable. I was taking part in a 48-hour sponsored bird race. Our challenge was to record as many species as possible in two days. All modes of transport were allowed and, as darkness fell on the first evening, four of us crammed into a tiny four-seater aircraft (plus one more for the pilot) and took off for the coast where we would search for seabirds at dawn the next day.

We were all pretty exhausted, and within 10 minutes most of us were asleep. Then suddenly we were startled from our slumbers by an almost deafening crack of thunder. Seconds later, the cabin and the sky around us were illuminated by sheet lightning, followed by another gargantuan explosion, and the sort of strobe lighting you are warned not to look at in nightclubs for fear it might trigger an epileptic fit. At the same time, the plane was rattling and rearing as if it were in the fist of King Kong. It was both terrifying and thrilling. People pay good money for this sort of ride at Disney World or Universal Studios. All it lacked was a computer-generated dragon to loom up in the windscreen.

Suddenly, there it was. Not a dragon. A bird. A big bird. For a few seconds, it flailed its raggedy wings as it nearly crashed into the windscreen. Then it veered away in a final flash of lightning.

'What was that?' I yelled. One of our Kenyan team members replied confidently: 'Crowned Eagle.'

'Have we had Crowned?' I asked.

'No. It's a new one!'

A cheer went up. The plane went down. Then steadied, and headed into the storm, and on to the Kenyan coast.

Balloons? They're for kids!

CHAPTER NINETEEN

The Call of Nature

A bogey bird is a bird you haven't seen but should have. A bogey bird may not be rare but it is elusive. For many years, my bogey bird was Capercaillie – the male is as big as a turkey, makes an unmistakable sound and its location is often well known or even signposted.

Such was the case in the Spey Valley in Scotland when Stephen Moss and I were working on an early episode of *Birding with Bill Oddie*. We endured a six-hour, decidedly unmagical mystery tour of the forest trails hoping for at least a glimpse of a Capper, which was 'definitely here somewhere'.

We found a feather and a small pile of probable poo, which was later reidentified as 'dog' by a man from the RSPB, who declared himself 'happy that we didn't see any Cappers, because that meant they were safe in the forest!'. We found it hard to share his happiness. Fortunately, another man from the RSPB had told us that if we had trouble – if! – there were two or three Capper cocks seen regularly near the Osprey Visitor Centre at Loch Garten.

'They are not actually visible from here,' explained the warden, as we poked telescopes and telephoto lenses through the viewing slots. 'But we have got a CCTV camera on the edge of the wood that's giving us great pictures.' To prove it, he switched on what I had assumed was a telly. Two male Cappers strutted across the screen and squared up to each other, like mythical creatures about to do battle. 'Is this a video?' I asked. 'No, no it's live. *Now.*' he emphasised proudly. 'They are only about 50 metres away, but they're behind the trees. You won't see them.'

'But I already have!' I countered. 'OK, they are on a screen, but they are live! So does that count?' I am sure the warden wanted me to 'lay my bogey', as it were, but rules is rules. 'You can't tick a bird on the telly. Unless you keep a telly list.'

'Which many birders do,' I reminded him. 'Maybe I should start a CCTV list?' But as if in protest, the screen went blank and the birds disappeared behind a snowstorm of interference. The picture returned, but they didn't.

More Capperless years went by, during which I clung to the consolation that one day we would return to Scotland and try again, but Stephen and I didn't like repeating ourselves. What we did enjoy was going somewhere different, for example, Sweden. We weren't even making a programme, but had been invited to witness and eulogise about the enormous spring gathering of cranes, of which the people are justly proud, and will be even prouder if the experience becomes a world-famous tourist attraction. It has all the ingredients: the birds look good and sound good, they are enormous, in huge numbers, spectacular as a flock and entertaining as family groups or in pairs, and they dance. Instant gratification guaranteed. Just what tourists need.

Stephen and I, however, had negotiated something a little more challenging. 'Have you had enough?' asked our Swedish host. 'We need to get you into the forest.'

The light was verging on crepuscular as we veered off the road and clattered along an ever narrowing forest track. Then we stopped. 'OK, you get out now. You get in the hide, quickly. I need to leave before the birds see us or they will not come. I shall be back in the morning. Maybe 10 o'clock. By then you will have seen the Capercaillies.' As he revved into a three-point turn he muttered something inaudible. Was it 'good luck'? Or 'fat chance'? The car's engine churred into the distance like a giant Nightjar, leaving us engulfed in a forest that was getting blacker every minute. The thick carpet of pine needles suffocated every sound. It was as silent as a graveyard. We shuffled towards our coffin.

The hide was neither cosy nor comfy. It was basically a large box with a narrow viewing slit at the front, and sackcloth curtains at the back, as the entrance and exit. We unrolled our sleeping bags, no doubt both deducing that a modicum of physical intimacy would be mathematically unavoidable. To minimise the risk, we constructed a low wall between the bags with our boots and rucksacks. Before long, darkness enveloped everything and eventually – after a few minutes of inevitable blokish banter – we both fell asleep.

We were woken not by the imperceptible brightening of the sky, but by a soft 'chucking' sound, not unlike water being poured from a bottle. We knew what it was. Every birder knows, whether they have heard it or not. The 'song' of the Capercaillie is one of the most surprising and comical songs in the bird world. The bottle-pouring sound isn't just repeated monotonously, it rises in speed and pitch and punctuates its climax with a resounding 'pop', similar to a champagne cork or a finger pulled out of a cheek. It would be silly enough if it were made by a small bird, but coming from an enormous plump Pickwickian creature it is as hilarious as it is unexpected.

It is not difficult to impersonate, so I did. Another bird answered. It wasn't far away. I contorted myself and ricked my neck peering out of the slot. My eyes climbed up a still unlit tree trunk. There – 9 or 10 metres up – hunched on a horizontal branch was my first unambiguous, in the flesh, not on a screen, male Capercaillie. At that moment, it was just a huge black ragged silhouette, but it wouldn't be long before it flopped down to the forest floor and engaged in some ritual sparring and popping with any other males in the area. All we had to do was to stay in our hide, keep still and quiet, and wait.

At which point, Stephen whispered in the sort of tone that normally indicates ensuing panic: 'Bill, I need to go!' He avoided my incredulous stare by crawling towards the exit. I verbally dragged him back. 'You can't!'

'But I need to.'

I was not sympathetic. 'Look, we've been in here – what? – 10 hours! We've only seen a silhouette. If you go outside, it'll be off! And it won't come back.' I could see that he was distressed. 'Just pee in the corner. It'll just soak in. I'll move my boots.' But he didn't move.

'That won't work,' he winced.

'Why not?'

'Because…' His hesitation spoke volumes. Talk about mixed emotions: annoyance, sympathy, confusion and a little bit of laughter!

If Stephen ever runs out of ideas for writing books or producing programmes – which he won't – there is a career for him as a commando. How he crawled through the sackcloth, did whatever was required, and crawled back again, I shall never know. I shall never know because I couldn't look for fear of laughing out loud and disturbing the Capper, which was still hunched up on its lofty perch. I was willing it to go back to sleep until Stephen had crawled back by my side. 'Please don't go. Please wait till it's light. Please find a rival and strut around going "pop".' It did. With two more. Which is why I now have to nominate another bogey bird.

Sorry? What do you mean I can't count it 'cos it was Swedish!

CHAPTER TWENTY

Black and White

A friend of mine works for the RSPB. One morning, he answered a call from a lady who asked him: 'I wonder if you could help me. I have just seen a strange bird in my garden.' If you are ever asked to identify a 'strange bird', there are two cardinal rules: first, assume it is a common species. Second, ask what it is doing. In this case, the lady offered that information anyway: 'It is eating the seeds on my birdtable. It's got a very pointy beak.'

'Ah yes,' responded my friend, as if he had already deduced or indeed anticipated what the bird was. 'What colour was it?' he asked. 'Er, sort of pale,' she replied. 'Pale brown? Pale grey?'

'Both really.'

'Buff? Pale buff, OK? And any other colours?'

'Yes, on its head, a bit of red, and I think a bit of black.'

It was time for the third cardinal rule: whatever the bird is, sound excited and share the joy. 'Oh, great,' he said. 'You've got a Goldfinch!' Her response surprised and perturbed him. 'It was definitely *not* a Goldfinch.'

'Is it still there?' he asked.

'I'll have to go and look through the kitchen window.'

There was a pause while she did. 'Yes, still there,' she announced. 'Still on the birdtable. Well, not actually on it. Standing by it. Pecking at the seed.'

Curioser and curioser! 'So, it's standing by the table and reaching up?'

'No, down. It's taller than the table.'

'What?! Heron size?'

'Maybe bigger. But sort of heron shape.'

'But it's definitely not a heron?'

'Of course not. I told you, it has a buffy body and red and black on its head.'

'And it's bigger than a heron. Blimey! It can't be!'

At which moment my friend remembered another cardinal rule: never make assumptions. The bird was not a Goldfinch. It was a crane!

The incident illustrates a couple of popular birdwatching adages: anything can turn up anywhere, and a rare bird is often a lost bird. One of the major factors is the weather, especially strong winds by which seabirds may be whisked inland and unceremoniously dumped somewhere bizarrely incongruous. For example, the Little Auk that made the front page of the *Birmingham Mail* by pitching into the 'Bullring' in the 1950s, when I was an embryo birder and the Bullring was a characterful open-air market, though still not typical Little Auk habitat. Little Auks belong in the far north of the Arctic, not in the English Midlands. They look like miniature penguins, but they can fly, and are prone to winter wandering round the North Sea, from which they may be plucked by easterly gales and end up 'wrecked', which is, by the way, the ornithological term for a bird being blown off-course, not for being sozzled on a Saturday night. Which takes me back to my friend at the RSPB. (No offence, you'll soon see the relevance.)

This time the distressed caller wasn't so much interested in identification as in emergency first aid. 'I have just found a bird lying on my patio. I think it may be injured. I've put it in a shoebox.' My friend applied the relevant rule: don't immediately say it is the RSPCA's problem (even though it is!), show compassion and concern. 'Oh dear, it's still alive is it?'

'It was when I last looked,' she said with no great conviction.

'That's good. In fact, that's excellent,' he added sounding almost excessively effusive. 'There have been a few Little Auk wrecks reported lately.'

'Oh, I don't know any of the birds,' interrupted the caller, 'except Blue Tits.'

'But this bird isn't blue is it?'

'No, it's black and white.'

'Good, good. And quite dumpy?'

'Well, yes. But quite little.'

'Exactly. Fantastic! I can drive over and pick it up, in about an hour.'

'Oh, that's so kind. Should I leave it in the shoebox?'

'Well, what it really needs is to get back to water. Tell you what... Run a cold bath, and put it in there.'

'Should I feed it?'

'Er, if you've got any tins of anchovies or sardines?'

'In tomato sauce?'

'No. I'll bring some fish. I'll be as quick as I can. And thank you *so* much for calling!'

Forty-five minutes later, they were climbing the stairs to the bathroom, discussing the patient's condition. The lady was clearly upset. 'It doesn't seem as lively as it was,' she whispered. My friend slipped into reassuring doctor mode, tinged with that tingle of anticipation every birdwatcher feels when they are on the brink of a close encounter with a rare bird. But as the bathroom door swung open, anticipation turned to disappointment, which turned to horror.

There, floating in the bath, was the bird. Black, white, soggy and obviously totally lifeless. She knew the answer, but she had to ask: 'Is it all right?'

'No. It is not what I hoped – I mean expected – it to be either. It is not actually a Little Auk. It is a Great Spotted Woodpecker!'

'A woodpecker? Shouldn't it be in a tree? Oh, poor little thing.'

'Yes, but it obviously wasn't well when you found it. I might have had to put it out of its misery anyway.'

The lady was not consoled. 'Maybe,' she snapped, 'but not by drowning it!'

CHAPTER TWENTY-ONE

Between the Devil and the BBC

In 2005 I was involved in a pair of programmes for BBC1 called *The Truth About Killer Dinosaurs*. As befits the grandiose title and the creatures themselves the production was on a pretty large scale. I was on the cover of the *Radio Times*, overacting between the jaws of a huge metallic model of a *T. rex*, every tooth as lethal as a javelin. Most of the show was filmed in an enormous studio that was also used to crash-test cars. We used it to crash-test dinosaurs, for example, one elaborate experiment involved propelling half a *Triceratops* on a motorised trolley straight into a brick wall until its skull shattered and its horns fell off, with a view to judging whether charging head to head with another dinosaur would have had the same effect. It would.

My personal favourite sequence was when we tested the clobbering effectiveness of an *Ankylosaurus*'s segmented club-ended tail by releasing a carefully calculated swipe at a frozen turkey. If you are having trouble envisaging that, think of the tail as a golf club and the turkey as a ball. Or the dinosaur as the equivalent of a medieval knight whacking people's heads off with his mace.

My personal close encounter with a killer was slightly less spectacular since it was achieved by the device of blue – or in this case green – screen, and involved me having to pretend to be terrified by two brooms leant against a kitchen chair, which – by the wonder of computer trickery – would eventually be replaced by an undeniably scary *T. rex*. When I saw the effect I was impressed, though I could barely resist shouting out 'It's not a killer dinosaur, it's a mop!'

Frankly, the technology made for tedious filming, but it was alleviated by gales of laughter, especially from one of our American palaeontologists whose job was to translate my commentary into a language that American viewers could understand! They say that Brits and Americans are divided by a common language. Maybe, but giggling sure brings us together! And believe me, playing 'whack the frozen turkey' is a real ice-breaker (pun slightly intended).

Talking of Americans, further proof of a generous budget and a major personal attraction was that the schedule included three days in Texas! One of the things I love about visiting different American states is that each one has its own clichés. No sooner had we left the airport than we were driving through oilfields, passing diners advertised by giant cowboys saying 'Howdy Pardner', and calling at gas stations where you could fill your tank for the price of a cup of tea. Best of all, the birds were Texan too. High in the glaringly blue sky, the 'buzzards' were circling just like they do in cowboy movies, though in fact they are not buzzards, they are Turkey Vultures. Whirring wings over the roadside meadows were – what else? – meadowlarks, and along the field margins were birds that looked like a cross between a lark and a sparrow. Lark Sparrow? Yep, you got it. Perched on wires and fence posts was a species I last saw in spring 1965 flitting around magnolia blossoms in Louisiana, and one that still gets my vote as one of the world's most elegant birds: Scissor-tailed Flycatcher. Even better, my eye was diverted by a creature scuttling along about 20 metres ahead of us as if daring us to catch up. 'A hare?'

suggested our driver. Nope. A Roadrunner. Just like in the cartoon, except they don't go 'beep beep'!

Pretty soon I could stand no more. Nature called. Our car stopped and I leapt out and disappeared behind a small clump of trees. 'How long will you be?' shouted our producer. 'Half an hour,' I called back, 'maybe a bit longer. Would an hour be OK?'

'Forty minutes. No more.'

It is, of course, not usual to haggle over how long an emergency stop is going to take, but I'm sure he'd realised that I was desperate to seek relief not from my bladder but from my frustration at being in a brand new location and not being able to do a bit of proper birding.

I was patiently stalking a Lark Sparrow – a new bird for me – when I was reminded of the reason we were in Texas. My binoculars followed the bird as it shuffled into what was clearly a completely dry riverbed. An ex-creek, I guess, in Texan. I noticed that what I'd assumed to be pebbles were all shell-shaped. Closer inspection revealed that this was because they *were* shells. Or rather they had been. A very, very long time ago. The more I looked, the more I found. Cockle fossils, mussel fossils, little fossils, big fossils. Very big fossils! You bet.

They call it Dinosaur Valley. There is a river, much wider than a creek, but especially after a dry period, the water can be very shallow or indeed absent. We set up the camera at the top of the bank, I looked down and we recorded my reaction. Wow! There below me, criss-crossing the riverbed, and as clear as if they had been stencilled overnight, were huge footprints. Two very obvious lines, and another area where the pattern was less orderly. We carried on down the slope and onto the shoreline, where we realised that what looked like sand was in fact hard flat rock. I waded into the shallows, and literally followed in the dinosaur's footsteps.

The true size of anything under water is, of course, hard to judge, but now I could measure each print in 'hands' (I needed more than two). Clearly, the producer's notion of me leaping from one print to another wasn't going to happen. An Olympic long jumper wouldn't have made it. Instead, I splashed around a bit and attempted to convey a word picture of what the palaeolithic scene might have looked like. This was obviously a drinking and bathing place. There were clearly different species visiting it, and experts have identified them. Some were relatively small, others were enormous. Some were harmless grazers, others fearsome carnivores, including the killer dinosaurs featured in our programmes.

We left the location astonished by what we had seen and delighted with what we had filmed. Incontrovertible evidence of life on ancient Earth, so different from now, and yet with many similarities to the modern world. Who among us would not be impressed? Quite a few it seems.

As we returned to the valley's main gateway, I veered towards a large white wooden-slatted building, which on the way in I had assumed was an information centre and gift shop. I was looking forward to buying fluffy dinosaurs for my granddaughters and a bag of assorted fossils for my grandson, and for me. However, as I approached the door I realised that it was not a shop; it was a church. The preacher and his congregation, far from celebrating dinosaurs, would be cursing them, if indeed they believed in their existence at all. Even more abhorrent was the concept of 'evolution', especially the suggestion that man himself is descended from 'lower forms', which is considered not merely a fallacy, but an iniquitous falsehood, a heresy and a sin. Such are the convictions and the creed of the 'Creationist' churches. There are an awful lot of them in the States, especially down south.

I have to admit that when you compare the two basic theories of the origins of life on Earth – either that it evolved over millions and millions of years or that God created it in six days – I personally find it hard to believe either of them. But Creationists v Evolutionists isn't just a matter of opinion or amiable rivalry. As you exit under the Dinosaur Valley Archway, the very first building is that church, challenging you to choose between confrontation and salvation. I could almost hear the preacher's voice: 'OK, you jest go ahead and look at them so-called din-ee-saw prints if you like, but after that, you'd better git in church and repent, for you have looked upon a lie! May the Lord forgive your sins.'

Sir David Attenborough told me that he was once on a lecture tour of the US, talking about prehistory and, of course, dinosaurs. Every show was a sell-out success – until he went down South. The hall was full, and the audience weren't actually hostile, but they were less than enthusiastic. When David had finished there wasn't the usual storm of applause. There was a mere trickle. Then everyone filed out in silence. David thanked them and withdrew to the wings where the theatre owner was waiting to escort him from the building, and possibly out of town. The man was uncomfortably silent, so David spoke: 'Er, was that all right? Only it didn't seem to go terribly well… did it?' The theatre owner took a deep breath and said – in that southern drawl that slips so easily from hospitality to menace – 'Folks round here don't care to think about that kind of thing.'

We couldn't possibly do a programme promising killer dinosaurs without featuring velociraptors. Perhaps the phrase 'The Truth About' implied that we were going to mollify – if not obliterate – their fearsome reputation, as established so chillingly in *Jurassic Park*. Well, maybe so. Scientific research has concluded that they were quite small and that they hunted in packs, probably seeking out prey that was already injured or sick. To demonstrate the relatively unthreatening nature and modest stature of *Velociraptor*, my producer had the quaint notion of comparing it to a turkey. Not totally inappropriate, since it is often mooted that birds evolved from dinosaurs. The idea was that I would crouch a little, thus approximating the size of a *Velociraptor*, with the similar-sized turkey standing beside me.

Our assistant producer was sent to procure a performing – or at least docile – turkey. There are Wild Turkeys wandering the woods in Texas, but they tend to betray their wildness by hiding or running away very fast (behaviour that has no doubt evolved as a response to being frequently shot at). The assistant producer phoned a turkey farm. The phone was answered by what sounded like a southern belle.

Farm owner: Hi y'all. How can we help you today?

Assistant producer: Er, we are looking for a turkey.

Owner: Well, mam, you sure come to the right place. Would you be wanting that plucked, or unplucked, or would you be fixin' to kill it youself? You can take 'em dead or alive!

Assistant producer: We want it alive. We want it for a television show.

Owner: A TV show? A show about turkeys! Well I never. You want a whole bunch of 'em? I can get you some real lively ones.

Assistant producer: No, we only want one. And we don't really want it lively.

Owner: And what kinda show is this?

Assistant producer: We're from the BBC.

Owner: The BBC! Oh my Lord! Hey Reuben, The BBC! *Doctor Who*, *Downtown Abbey*, Benny Hill! Oh my, we *love* the BBC. Oh my my. One of our turkeys gonna be on the BBC. Praise the Lord. We are so honoured! Tell me, mam, what kinda show is this?

Assistant producer: It's about dinosaurs.

There was a moment's silence as her voice snapped into menace mode.

Owner: Listen, lady, I don't care who you from. We are not letting one of our turkeys appear on no programme about that e-vo-lution shit! No, mam.

Fortunately, we managed to get a Buddhist turkey flown in from Florida.

PS *The Truth About Killer Dinosaurs* was issued as a BBC DVD in 2005, and may still be available somewhere!

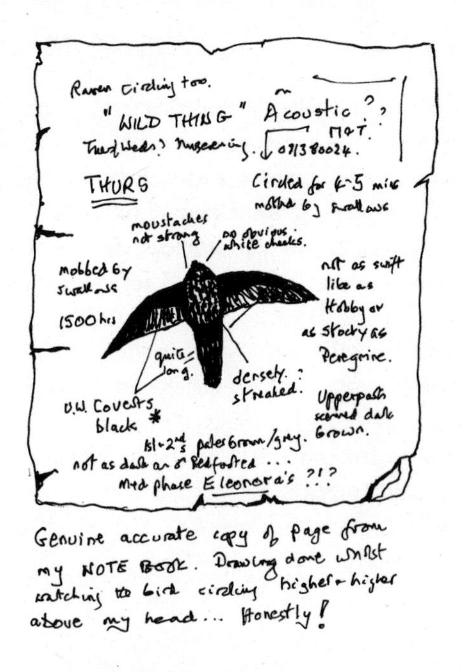

CHAPTER TWENTY-TWO

One That Got Away

A birder friend of mine, while admitting that he had very little artistic ability, nevertheless reckoned that if he encountered an unfamiliar bird in the field, even the most rudimentary thumbnail sketch was worth more than several pages of notes, at which point he showed me his hasty pencil portrait of an as-yet-unidentified bunting. I commented that it certainly looked more like a thumbnail than a bird, which was only as unkind as it was true! Nevertheless, this certainly has not deterred him from sketching before writing and – for several reasons – I have to agree that that is the best way.

For a start, scribbling notes largely involves looking down at the page, not up at the bird. Indeed, there is a danger that concentrating on meticulous details of supercilia, tertials, primary tips, crown stripes and rumps can be so absorbing that by the time you have finally decided to scope it the bird has flown off when you weren't looking. Mind you, I can testify to the fact that there are some 'creative' twitchers who won't let a little thing like a fleeting or inadequate view of a 'possible' rarity stop them producing a two-page spread of field sketches from all angles and in a variety of poses, quite likely with little arrows indicating the diagnostic features. This, however, is not proof that they actually saw diagnostic details, but only that they know what they are. Whatever the truth, they are convincingly illustrated in their drawings, suitably labelled: 'Tail spread to reveal white outer tail feathers. Showed grey rump clearly on landing. Rufus streaking visible on flanks when bird is perching.' Which, of course, it is doing in the picture. By the way, I do hope you don't think I am insinuating there is such a thing as 'illustrated stringing'. I would prefer to call it 'enhanced field sketches', which as it happens is my personal favourite genre of bird art.

Some people sketch what they see. Others sketch what they want to see. Or perhaps wish they'd seen. I myself belong to the second camp, but only when I fancy doing a 'proper painting'. When faced with a suspected or even certain rarity I am with Mr Thumbnail. I draw – or rather scribble – while still looking at the bird through binoculars or telescope. What I produce is a tad more avian than a thumbnail, but it is barely recognisable as a bird. Ear coverts end up where undertail coverts should be, and the wing (no point in depicting two 'cos they're both the same!) may be completely detached. The overall result looks more like one of those puzzles in kids' comics, where you have a few apparently random lines, but when you join up the little numbers or dots it becomes something recognisable (though probably not a rare bunting).

So, I confess, my field sketches are a mess, but they do their job, which is to record enough detail to identify the bird. Once I have the identification sorted, I may attempt art. In fact, I only ever paint on holiday, so most of the species are Mediterranean rather than British. Golden Orioles, Woodchat Shrikes, Hoopoes, Azure-winged Magpies, Red-necked Nightjars and feral waxbills, from which miscellany you may deduce that I am a regular to southern Portugal. I am so familiar with the birds there that I feel I don't need them to pose. Instead, I set myself up in 'watercolour corner' (other family members paint other things) where I do a fairly rapid pencil sketch of my chosen bird or birds and colour in a few bits. It is all a bit cursory, because I am impatient to get on to the fun part, which I call 'designing the set and providing the props'. I rarely show a bird perched, feeding or doing whatever it is doing in the same context I saw it originally. For example, I saw the Woodchat on a wooden fence. I put it perching on barbed wire, with a few tufts of sheep's wool and a discarded fag packet snared on one rung and a half-eaten lizard on another. I also added a broken wine bottle – Mateus rosé – overgrown by thistles at the foot of the fence post. Some of my pictures look as though they were designed for a 'Take Your Litter Home' campaign. In fact, they often involve me taking other people's litter home. It is not easy to depict a Portuguese ice-cream wrapper from memory. By the end of the holiday, 'watercolour corner' is scattered with more detritus than a teenager's bedroom. I have considered entering it for the Turner Prize. With no due modesty, I admit I am proud enough of my little paintings to not mind people seeing them, or even bidding for them in a charity raffle. However, the same definitely does not apply to my field sketches.

Ironically, I may well have been intimidated by my first 'mentor' – the wondrous Dr E. A. R. (Eric) Ennion who was in charge of Monks' House Bird Observatory back in the 1950s. Many times as a teenager I sat with him on a Northumberland beach gazing out at the Farne Islands, while he dashed off instant drawings on the back of a fag packet, which he usually gave me to keep. Inevitably, alas, I didn't. The back of a fag packet was also the canvas of choice for another legendary field artist, R. A. Richardson, the Prince of Cley. He wasn't really a prince, of course, but he did habitually wear a suit of shining black leather and ride a proud and noble steed. A Harley Davidson I presume. As a teenager, I sat at the feet – or rather peered over the shoulder – of these two talented and incredibly nice men as they illustrated whatever birds we had been discussing or were watching. They were both generous and entertaining people as well as genius artists. If I had to establish a preference, as they say, I think I would have to favour Ennion, whose finished paintings somehow managed to retain the immediacy, movement and life of field sketches, despite his birds having jagged outlines that verged on impressionistic. It's a style that flourishes in the work of John Busby.

So, whether your motive is art or identification, field sketching is a talent to be envied and a skill to be learnt. If nothing else, if you are fortunate enough to find a rarity, a decent sketch, done while the bird was visible, will surely impress those who carry the poisoned chalice of judgement, such as the Rarities Committee. On the other hand, thanks to ever-evolving technology, it is now possible to capture a digital image on just about any implement from a phone to a 'tablet'. Even a camera! So ubiquitous are these devices that some rarities committees have announced that they will no longer accept any claims that don't have photographic evidence. Of course, in most cases, there is. But if there isn't, what about a decent, accurate sketch?

Do you sense a story of string or injustice about to unfold? The date was mid-September, a few years ago. The place: Tresco on the Isles of Scilly. It is early afternoon and it is warm and sunny. I am lounging almost on my back in our cottage garden. Hirundines are circling lazily on the thermals, along with a few gulls and the occasional raptor. One in particular catches my eye and makes me grab my binoculars. It is a falcon. Kestrel? No. Hobby? No. Peregrine? No. Merlin? Definitely not. That accounts for all the species I had seen on the island so far that week. This bird wasn't any of those.

The truth is that I knew what it was but I would need to prove it. I rapidly did a sketch, as it circled higher and higher and then drifted out of sight. As I scribbled a few notes, I found myself muttering: 'Black underwing-coverts. It's the only one.' All that remained was to flick to the falcons in a European field guide where I would surely confirm what I was actually sure of. However – I know it seems a bit arrogant – but I don't bother to take a field guide on a British holiday. Besides, there is usually one among the books in the accommodation, alongside the Ian Rankins and the various Shades of Grey.

However, the only bird book on our shelf was purely British and belonged back in the era of 'I-Spy'. Off I scampered towards the quay, keeping one eye on the sky in case the bird reappeared, while thinking: 'Perhaps I should get one of those new-fangled smartphones.' With my other eye, I spied two island kids kicking a ball around their garden. I assumed they were too young to know much about European raptors, so I asked: 'Is your mum in?' At which moment, mum appeared. On the mainland, she would no doubt have hustled the kids away from the strange man who was suspiciously panting with excitement. Instead, I simply got a typical Scilly greeting: 'Lovely day isn't it?' I agreed effusively, and asked: 'You don't happen to have a bird book do you?'

'Oh, I'm not sure,' she replied. 'I think we have. Do you want to come in and have a look?' That wouldn't happen on the mainland either.

Thirty seconds later, I was flicking through the pages of a relatively recent *Birds of Britain and Europe*. After another 10 seconds I was perusing a plateful of falcons and smiling at the first line of the text: 'The only European falcon with black underwing-coverts.' Eleonora's Falcon. Of course, I knew that, didn't I?

I telephoned the appropriate resident Scilly birder, knowing that he would spread the news. Next morning, I got a call myself. Somebody had reported what 'might have been' my bird over another island. There was also a rumour of an Eleonora's 'somewhere in Cornwall'. To cut a long, rather annoying, story short, in due course I submitted my claim, and a copy of my field sketch and scribbles. I added that I knew the species well from Majorca. Eventually, I got a reply, and a verdict. The letter stated that 'there seems to have been some confusion as to the age of this bird.' I had diagnosed it as a first-year male. It certainly wasn't a totally black adult. After looking at pages of photographs in specialist raptor books, I concluded that it might have been an intermediate morph or a dark female. Eleonora's Falcons notoriously come in various shades! But there is one feature that they all have and other species don't – black underwing-coverts!

My bird was rejected. I was dejected. But not corrected! I don't know what those other people saw, or what they said, but I know what I saw, and what I drew. Mind you, I have occasionally wondered: did those who passed judgement even look at my drawing? And another thing: if it had been drawn by Eric Ennion or Richard Richardson, or Killian Mullarney or Chris Rose, would they have accepted it then? Or would they have demanded a digital photo?

CHAPTER TWENTY-THREE

Invisible Bird

Some birds are hardly worth seeing. Their looks aren't what make them special. Which in some cases is just as well, since it is almost impossible to get even a fleeting glimpse of them. I rather feel that about Nightingales, and Cetti's or Grasshopper Warblers, and certainly about Quails. Nevertheless, they are all shameless show-offs compared with the Noisy Scrub-bird, an extremely rare Australian species. It is well named, in that its song is so loud that it can carry over a kilometre, and it does indeed live in 'scrub', more specifically in a small area of scrub at the very southern tip of Western Australia. However, it also qualifies for some kind of alternative name such as the 'might as well be invisible scrub bird'. This would also act as a warning to those birders intent on adding it to their list, such as, for example, myself. It must be a dozen or so years ago that I drove down from Perth to Two Peoples Bay with a trio of eminent Aussie 'birdos'. We say 'er', they say 'o'! We all say 'twitching'.

Since these fellas were part of a Save-the-Noisy-Scrub-bird-type project, they knew instantly which sections of the considerable expanse of thick vegetation held territory-holding males. Not that I wouldn't have figured it out for myself, when my ears were serenaded by a burst of birdsong that sounded rather like a Cetti's Warbler using a megaphone. 'Fruity' and 'explosive' would be apt adjectives, and of course 'loud', or indeed 'noisy'. A few seconds later, another one replied that was surely using a microphone and an amplifier. Then another song at even more volume. He had obviously turned his amp up to 11. My Aussie mates were clearly enjoying my slightly bewildered reactions. 'How many of them are there?' I asked. They smiled: 'Just one!'

Noisy Scrub-birds are real little teasers. Their song attracts your attention and lures you into the bush, much as sirens lure sailors onto the rocks. You creep forward, homing in on the dense and shady spot wherein surely lurks the hidden singer. Then there is a slightly longer silence, broken by a burst of song, further away and from a completely different direction. You state the obvious: 'It's moved!' Only to be corrected: 'Not necessarily. They are ventriloquists!'

'Oh. Have they got their dummies?' I quipped.

'You're the dummy, mate! No offence, but you'll never find them by following them. You've got to get them to follow you.' I guessed what was coming next.

Nowadays, it would be an MP3 player or a smartphone. Back then it would have been a portable cassette tape recorder. However, being a bit of a purist, I wasn't totally comfortable with 'tape-luring'. I always had a vision of the bird being really narked when it realised it was being flirted with or challenged by a machine. My compromise was to produce my Audubon bird caller, which is a small device that looks like a little wooden whistle, but is in fact 'played' by twisting it in your fingers so that the friction produces various squeaking noises that supposedly puzzle small birds enough to make them come closer to investigate. It's the equivalent of 'pishing', but without tingly lips. Neither technique is guaranteed to always work, but this scrub-bird found the Audubon squeaker irresistible. It ceased flaunting its ventriloquial skills, and its song got noisier and noisier and closer and closer, until I was convinced that the bird had to be in the middle of a small isolated bush, no more than a metre away. Even the Aussies were cautiously positive. Whispering ensued, and we literally surrounded the bird, so that it was equidistant from four keen-eyed, sharp-eared birders, armed with close-focus binoculars. We waited. Then it went quiet, or rather, it ceased to be noisy. Then it sang again, from what sounded like 100 metres away, but for all we knew it might as well have been under my boot. But when I lifted my foot, it wasn't there.

I have still not seen a Noisy Scrub-bird. Well, actually I did see one released – launched more like! – back into the wild from a 'captive-breeding' aviary. I saw it for barely three seconds, and only in flight, but well enough to confirm that it was yet another little brown bird. Hardly worth seeing? Well, that depends… can I count it?

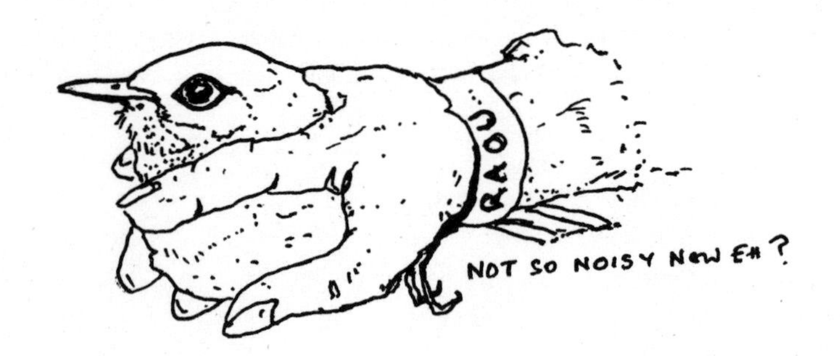

CURIOUSER AND CURIOUSER

CHAPTER TWENTY-FOUR

In the S**t

Back in the 1960s there was nothing I looked forward to more than a day out at a sewage farm. Young people today wouldn't know what a sewage farm was. 'They used to grow sewage? And harvest it?' No, they pumped it into lagoons. 'Lagoons of...?' Quite. 'Didn't it smell?' Oh yes. We always used to check the wind direction before we walked round. The birds loved it. In 1970, Wisbech Sewage Farm was declared the best place for migrant waders in Britain, if not in Europe.

Sewage is, of course, universal. Twenty-nine years ago I did my first proper wildlife series for the BBC. The producer's idea was to show what it was like for a British birder to be plonked down in a totally unfamiliar country. For example, Bill Oddie in Papua New Guinea. 'So, Bill,' he asked me, 'how would you go about finding the best birding places?' My reply would have been the same anywhere in the world: 'I'd ask if there was a local sewage farm.'

As it turned out, there was a beauty, just outside the capital city, Port Moresby. Locals rather poetically called it the 'S**t Pit'. As at all the best sewage farms, there were not only lagoons, but also reedbeds, ditches, damp grassland, hedges and trees, sprinklers and – at this one – a network of neatly trimmed grassy banks you could stroll along, while admiring relaxing ranks of wildfowl, egrets, herons, gulls and terns. Easy birding, and very easy filming. Unlike the rest of the trip, most of which was spent sloshing around in damp and gloomy rainforest, failing to find whatever we were looking for. I still smile at the irony of doing a series enticingly entitled *Oddie in Paradise*, which was mainly shot at the local sewage farm. A shame the producer didn't let me say so! Alas, since then, the S**t Pit has become the haunt of thieves and muggers and declared a no-go area.

However, there are no such dangers or delinquency at what may well be the best and possibly biggest sewage farm in the world, at Werribee in south-east Australia. Not only does it deal with the effluent of more than half of Melbourne, but it is also home to literally tens of thousands of birds, of nearly 300 species, including a colony of highly endangered Orange-bellied Parrots, which, by the way, do not mingle on the mud with the waders and wildfowl, but stick to the more manicured lawns and plantations.

Werribee Water Treatment Works (it sounds like a spa, but it's still a sewage farm!) is a king-size example of what Richard Mabey called 'unofficial countryside'. I'd amend that to 'incidental nature reserves'. Both phrases imply precarious impermanence, and to prove it Wisbech has been 'modernised', as has the nearby Cambridge Sewage Farm, another star venue in times gone by. Most others have gone the same way.

There is, however, one whose state remains a mystery. Perry Oaks Sewage Farm was too close to London's Heathrow Airport to be anything but an ear-shattering day out, but it was worth it. The birds ignored the planes, as did the birdwatchers – including myself – who slipped through, or even over, the security fence and scuttled around under cover of the embankments, occasionally popping up to scan a lagoon, flushing snipe and assorted shanks, plovers and sandpipers. I never did know whether or not Perry Oaks was private, but a terrorist threat to shoot down a British airliner assured that it definitely was. Thus, one of London's birding hotspots became totally inaccessible. It presumably still attracted birds, but no one was allowed to look at them.

Then, as the terrorist alert abated, another threat arose: Terminal 5! I am sure there were lots of objections to Terminal 5, but I doubt anyone pleaded that it would disturb or entirely obliterate Perry Oaks Sewage Farm. The truth is, I don't know if it did or not. Is there still a totally unwatched 'incidental nature reserve' down there? Or did they simply bury the sludge beds under the concrete of Terminal 5? Do you remember the film *Poltergeist* where a development had been built on the site of a cemetery, but 'they only moved the headstones!' thus desecrating the departed, whose spirits wreaked a terrible revenge?

Well, call me crazy, but every time I fly from Terminal 5, I swear I can hear the plaintive calls of Ringed Plover and Ruff, and I begin to feel a powerful force is trying to suck me down, down into a lagoon of... Now that's what I call a horror story.

CHAPTER TWENTY-FIVE

Jewels in the Dark

I found an old lady in my lampshade last night
Of course she was dead, she'd been lured by the light
But her lace was unsinged, and her smock undefiled,
So I took her downstairs, and I called to my child…

The Old Lady was, of course, a moth. The child was my youngest daughter, Rosie. 'Why do you think it is called an Old Lady?' I asked. 'Because it looks like a lacy frock thingy.'

'A crinoline.'

'Whatever.' After a few more seconds of moth perusal, Rosie and I spoke in unison: 'And this affects me how?' This immortal expression of teenage indifference was first uttered in response to me showing Rosie a fox when she was about 13. She is now 24, and while she has hardly blossomed into a committed naturalist, she is appropriately appreciative of most of the wild things I draw her attention to. Meanwhile, the feigned disdain of 'and this affects me how?' has become a sort of family catchphrase. As indeed has the outraged cry of 'Dad! Oh my God!' indicating that someone has opened the fridge door and discovered that the egg rack's eggs have been replaced with small specimen jars. They may look like moth coffins, but I assure you the occupants are not deceased. They are sleeping, or rather in a torpor. When I release them in the garden their wings will start to shiver and quiver, until they fly off, hopefully avoiding the nimble and intelligent Robin who has learnt to recognise the appearance of my moth tray, moth book and moth jars as the promise of an easy meal.

I am an inveterate garden moth-er. This year it was 25 March when I first set up my trap. It is very basic. A wooden box, an especially bright light, two sloping sheets of clear plastic and – providing shade and warmth for anything that is lured in – a layer of egg boxes. I like to think that this means the moths are not so disorientated in the fridge. March 25th had been a record-breakingly warm day, but the night was chilly, so I wasn't too optimistic. Nevertheless, the morning inspection of the egg boxes did not draw a blank. It produced five moths of three species. Two Common Quakers, two Hebrew Characters and a Brindled Beauty. Add in the Old Lady, and you may deduce one of the reasons I love moths. Those names! Some, simply descriptive: Scalloped Hook-tip, Purple Treble-bar, Pinion-streaked Snout; others more playful: Maiden's Blush, True Lover's Knot; and a few almost fearsome: Death's-head Hawkmoth. All of them are testimony to the assiduousness and imagination of the Victorian naturalists who catalogued and named them.

Get into moths and you instantly become a member of a club. It may be literal – 'moth groups' abound – or it may seem more subversive or even slightly secretive. I first realised this when I filmed an item for *Bill Oddie Goes Wild*. My producer, Alex, had clearly grasped the atmosphere, and was determined to relish it. We convened at dusk and formed a circle round a single lamp while chanting: 'The first rule of Moth Club is you don't talk about Moth Club.' Then each member solemnly intoned their name and origin: 'Jack from Tewkesbury, Timothy from Chipping Norton, Bill from Hampstead', and each one stood by their traps, which incidentally were all home-made and collectively looked like an exhibition of outdoor scrap sculptures. There was a ritual synchronised turning on of mercury vapour lamps, followed by two or three hours of me talking to the camera and the club members quaffing red wine, an activity that continued well into the small hours as we all pored over identification books. Another rule of Moth Club: 'No moth goes unnamed.'

Alex's closing shot through the cottage window was of us crouched over a desk surrounded by wine glasses and empty bottles, all romantically lit by an oil lamp. You can just hear the lady of the house offering a choice of coffee or cocoa, while inebriated voices struggle with tongue-twisters like a Three-humped Prominent or a Triple-spotted Pug. At least one person had fallen asleep. It was one of those experiences that remind me that there is more to watching wildlife than the wildlife. Company, humour and shared enjoyment.

CHAPTER TWENTY-SIX

On the Fairway

Birdies, eagles, albatrosses. Am I talking wildlife or golf? Well, both actually.

The fact is that most naturalists – and birdwatchers in particular – are rather fond of golf courses, especially if they are sited by the sea, which many of them are. It means that many migratory birds' first landfall is on a golf course. The greens and tees may be a bit too velvety, manicured and possibly doused in insecticide, but the short-cropped fairways are much beloved of wheatears, pipits and larks, while warblers and thrushes lurk in ditches and any available scrub or bushes. What's more, landing in the 'rough' may be a nightmare for golfers but it is a joy not only for birds but also for small mammals, wildflowers and attendant insects, especially butterflies. Indeed, there are some golf courses that look as if they have been designed for nature. Some of them have.

A few years ago I was given a guided tour by the environmental manager of a holiday village in Suffolk. He showed me an area destined to become a golf course, but also a nature reserve. Islands of 'rough' had been left undisturbed and – encouraged by a little planting and management – were glowing on that May morning with heath-loving wildflowers, many of which had lain dormant until the construction work had emancipated them. These included some of the common orchids. If you hanker after less common varieties, there are few better sites than Sandwich Bay in Kent, or – to be more specific – the fringes of the Royal St George's Golf Course, sporadically home of the British Open and permanently home of Lizard Orchids.

So, is there a Ryder Cup of nature-friendly courses? Well, I have to admit that my most lavish experience of 'life on the links' – surely the title of a TV programme? – was in the US, at Walt Disney World, in Florida. There is a lot more to the Disney complex than Mickey Mouse, Thunder Mountain and Cinderella's Palace. An area almost the size of an English county is covered by woods and waterways, much appreciated by the usual brazenly tame Florida birds, such as herons, darters and ibises, plus the occasional alligator, and three kinds of chipmunks: real, animated and actors in furry suits.

There are also many golf courses. One of them became my temporary local patch for the week. I saw warblers and woodpeckers in the surrounding trees, kingfishers hovering and diving into what I assumed was a 'water hazard', but was also an ideal nesting beach for a small colony of Least Terns. I wondered if they ever ended up incubating a stray ball instead of their egg. Most fortuitous of all, it being a baking late August, the drought was drenched every morning by a cascade of sprinklers that left the greens and fairways soggy and soft and an ideal habitat for migrant waders, the passage of which varied entertainingly from day to day. When the golfers arrived the waders simply fluttered a few yards to another mini-marsh, both birds and golfers oblivious to the danger of a hooked tee shot bringing down a Semipalmated Sandpiper. At which point surely no golfer could possibly resist announcing: 'I got a birdie at the 12th!'

I see no reason why wildlife and golf courses should not co-exist. Nor, indeed, birders and golfers. Unless, of course, there is a rarity involved. This happened back in the 1980s in south Wales.

The bird in question was a Little Whimbrel (a Whimbrel is like a Curlew only smaller, a Little Whimbrel is even smaller) normally found in Australasia, and only twice seen in Britain. That is rare. It was first spotted within the bounds of Kenfig National Nature Reserve. However, just before dark, it flew to roost on the adjacent golf course. The news spread like wildfire. All night the twitchers of Britain travelled from far and wide. Before first light, several hundred of them were assembled – not at Kenfig Nature Reserve, where they were welcome, but at the end of the first fairway, where they weren't. Imagine the scene as, just after sunrise, the first golfer shuffled to the first tee, placed his ball on the pin, selected his driver, adopted his stance and peered down the fairway. At which point, 500 khaki-clad twitchers appeared over the horizen, plonked down their tripods, and pointed binoculars, cameras and bins apparently straight at him! Even Rory McIlroy would have sliced his shot.

'Eagle', 'albatross', why not 'little whimbrel'? It never has caught on.

CHAPTER TWENTY-SEVEN

Birth of an Island

Volcanoes are best viewed from a distance. This is not just a matter of health and safety, it is only from far away that you can appreciate the classic volcano shape of a massive inverted ice-cream cone, usually with the summit shrouded in cloud, which is quite reassuring if you had been considering a long trek to the top, because by the time you got there chances are you wouldn't have been able to see a flipping thing. I speak from experience.

If the volcano was active, you probably wouldn't want – or indeed be allowed – to approach the crater (TV presenters and film crews excepted, of course) which is fine by me, since the cosmic firework display of molten lava, elastic rock and cascades of fiery sparks looks so much more fun when you are gaping at it from way below, rather than fleeing for your life.

Eventually, of course, most volcanoes run out of steam, smoke and fire, and become habitable by wildlife, especially birds, some of which evolve to become either a new species or at least a volcanic 'race'. This often involves matching their plumage to the colour of their habitat in the interests of camouflage. Old, cold lava is, of course, black. This has led to some contradictions in our expectations of what colour some birds 'should' be. For example, the Seychelles has a 'black parrot' which, I'd say, contravenes the Trade Descriptions Act. The Galápagos have a Lava Heron and a Lava Gull, and a cormorant which is a common cormorant colour – black – but has also lost the power of flight. In case you are thinking that all this is the result of insular isolation, take a coach ride to the crater of the volcano just outside San José in Costa Rica and there you will find the Volcano Junco and the Sooty Robin, if you can spot them. Dark grey birds among dark grey rocks are not conspicuous.

Of course, evolution is a slow process. We can't literally watch it happening. However, 30km off the south coast of Iceland there is a sort of small-scale preview. The island of Surtsey rose from the sea in November 1963. It is therefore approaching its 50th birthday. Indeed, a celebratory conference is going on in Reykjavík even as I write. After eons of volcanic pregnancy, the birth began with several days of underwater contractions, followed by a spectacular eruption of water and smoke, which eventually cleared to reveal a newborn island. This provided a unique opportunity to observe its development, especially the arrival of life.

The only human beings allowed on Surtsey are 'licensed' scientists, and the BBC. Well, my film crew and I were. For a couple of hours, about 10 years ago. We were dropped by helicopter and left to tiptoe on hallowed ground. It wasn't spectacular. A few wisps drifted unthreateningly out of the crater, below which were slopes of slightly warm volcanic dust, as black and as powdery as pepper. At a glance they seemed lifeless, but then I realised they were speckled with tiny pinpricks of green. Seedlings. Not planted, but borne on the wind or on the feathers of birds. I sank to my knees to get a closer look, but it must've looked as if I was worshipping the advent of life itself.

My whimsy was interrupted by a bird zipping past me. A single Swallow, intent on becoming a pioneer perhaps? Or prospecting for a new nesting place, perhaps under the eaves of the scientists' hut? But he mustn't have liked it. Or maybe he couldn't find an equally adventurous mate. A recent survey of Surtsey's breeding birds records only 13 species. Swallow is not one of them. Most are seabirds.

The most numerous land bird – 12 pairs – is the Snow Bunting. There's an isolated population, living somewhat incongruously in a land of lava. Surely rife for a spot of evolution? Imagine the news: 'Surtsey's Snow Buntings turn black!' Darwin would love it.

first the black head, then the rest?

CHAPTER TWENTY-EIGHT

Hearing Test

Some years ago, what I would have called an 'elderly gentleman'– until I became one myself – hailed me on Hampstead Heath and asked: 'Where have all the Treecreepers gone?' Years of being asked this sort of thing have taught me that it is impolite to respond with: 'There's one just flown past you and landed on that tree trunk,' because it could be interpreted as: 'It's not the Treecreepers that have gone, it's your eyesight!' Instead, with the blasé swagger of a relatively compos mentis 60 year old, I simply replied: 'Well, actually, I saw a Treecreeper quite recently,' resisting the urge to add, 'in fact, I am still looking at it!' I simply bade him an encouraging 'good luck' and pottered off chuckling to myself and muttering: 'You should've gone to Specsavers!'

At which point, I heard the Treecreeper start to sing. But the gentleman didn't. It was at that moment it dawned on me that I should've asked him: 'How's your ears?' I thought of calling back to him, but I didn't. I was sure he wouldn't have heard me.

The truth is that what we call 'birdwatching' could just as appropriately be called 'bird listening'. Take a walk in the woods, and you will certainly hear more than you will see. Added to which, it is almost always a call or song that first attracts your attention. Treecreeper is a perfect example. They do exactly what it says on the tin: they creep up trees. A bark-coloured bird on bark. Unless and until they move, they are nigh on invisible, no matter how good your eyes. However, they have a sharp-edged 'peepy' little call, and a wispy down-the-scale song, with a debonair little flourish at the end. It is unmistakable, *if* you can hear it. The problem is that the call and song are very high-pitched – almost up in the range that would annoy bats – and human hearing almost invariably deteriorates from the top. Some people can't hear Goldcrests. Others can't hear Grasshopper Warblers. I once did a sponsored birdwatch with a friend who was losing both his hearing and his mobility. We had to carry him to within a few metres of a reeling 'gropper' which the youngest ears had detected half a mile away. His cry of 'I can hear it!' seemed genuine, but surprised us, because by then we couldn't! Never mind, it was for charity. The gentleman on the Heath I mentally christened 'The man who can't hear Treecreepers.' 'I wonder if he knows,' I mused. 'Shall I tell him? No. Let's face it, it'll come to us all.'

My personal moment of truth occurred about a year ago, in early spring, on top of Parliament Hill, which is generally recognised as one of the best places in London for 'viz mig' – visible migration. Some mornings, hundreds or even thousands of birds fly over. I – often accompanied by one or two other viz-mig aficionados – will stand on the hill for hours gazing at the skies and yelling out bird names. Members of the public probably think we belong to some weird esoteric cult, which in a way we do. Even our language is unintelligible to the uninitiated. 'Ten chaffs going north, four goldies, reed bunt west' (that's Chaffinches, Goldfinches and Reed Bunting). At this time of year there will also be many a cry of 'mipit!'

Meadow Pipits are small streaky brown birds that to me are the true harbingers of spring. More than Cuckoo, Swallow or the first warblers. Most British Meadow Pipits breed not so much in meadows as on moors. Tramp through the heather anywhere from Cornwall to Shetland and you will probably kick out a few 'mipits'.

However, a large percentage of the UK population winter in Spain and southern Europe. In early spring, they come flooding back. Some of them pass over Parliament Hill, flying fast and high.

Some of them are so high that they become 'inviz mig', but we know they're there, because they call. It is not an impressive sound. More of a short squeak, rendered phonetically in the books as *tsip*, but the consolation is that they call a lot. In fact, I don't think they are capable of flight without calling. A silent Meadow Pipit has surely lost its voice.

However, one morning in March 2011 my birdwatching chum Dave must have thought *I* had lost my voice. We were on the hill staring upwards. Every now and then, Dave called 'mipit!'. I didn't. I was silent, and – to me – so were the birds. I managed to spot a couple of dots at the sort of height and distance that everything looks much the same. 'Those two, Dave?'

'Mipits.'

'Did they call?'

'Yes. Several times. Couldn't you hear them?' I felt blighted, embarrassed and depressed. The answer was 'no!'

I am 70. I still get out on the Heath, and I still enjoy 'viz mig'. I can still pick up the flight calls of chaffs, goldies and reed bunts. But, just as my hearing is not what it was, neither am I. I am no longer 'the man who can identify all the bird calls', I have become 'the man who can't hear mipits!'

CHAPTER TWENTY-NINE

Going Overboard

Some years ago I was offered the job of presenting *Breakaway* for BBC Radio. I accepted, assuming that it would be tantamount to a world tour subsidised by your licence fees. As it turned out, I never got further than Tewkesbury. I reminded David, the producer, that he had promised me one proper 'big trip'. Naturally, he honoured his promise. The trip was big in distance but not in length. Costa Rica for three days!

We flew to San José (presumably the same one that Dionne Warwick asked the way to). The first morning, we took a taxi up the local volcano. The higher we drove, the thicker the fog got, until we arrived at the edge of the crater which was totally invisible. 'Good job it's radio,' I muttered, and duly recorded my reactions to an awesome landscape that I couldn't actually see. We had our first sequence.

For the following day, David had planned an excursion that was guaranteed to be much noisier and have the sort of unpredictable excitement that comes over so well on radio. The taxi dropped us off on a bridge. We were instantly aware of the burbling and roaring of rushing water. Below us, a river – swollen by recent deluges – was in turmoil. It sounded great, but looked fearsome, especially as we were due to go white-water rafting. My feeble plea of: 'Can't I just act it?' was rejected, and I was squeezed into a wetsuit, along with David and half a dozen others – we had a suit each of course. So did an attractive Amazonian blonde American girl who stood astride the prow of our inflatable rubber dinghy and took us through some synchronised paddle exercises. 'When I say "right", you guys stop paddling. When I say "left", you guys…'

'Stop paddling?' I pre-empted.

'You got it!' she enthused. For 10 minutes we practised, until we were as slickly drilled as a varsity boat-race crew. Then we lowered the dinghy into the water.

There was instant mayhem, as the current swept us downriver, dipping, tipping, swirling and buffeting. Paddle drill went haywire. The Amazon barked: 'Left, right, left, *right*!' She might as well have yelled: 'Everybody panic!' David was attempting to shove a microphone in my direction, while protecting his precious tape recorder from the spray. He began to interview me. 'Apparently it's really good for wildlife along the river. Tell us what you can see.' I was on the brink of blasphemy, when the Amazon called our attention.

'Hold tight everybody (like we needed to be told)! Round this corner yesterday there was a big boulder which could… Oh my God! It's gone!' Judging from her tone, this was not good news. Instead of a mighty boulder, there was now a cavernous hole, down which the river was writhing like a maelstrom, as if Costa Rica itself was going down the plughole. The boat tipped sidewise. I fell out.

I remember being under water. I remember reaching upwards and a hand grabbing mine, and I remember being hauled back on board like a harpooned seal. I may have blacked out. I may have fainted. How long I was 'gone' I shall never know. The next thing I was aware of, I was kneeling at the prow proudly paddling along Red Indian style. The river was now placid, and there was indeed wildlife along the banks. But I was oblivious. My eyes were glazed. David asked me: 'Are you all right?'

'No,' I replied.

'Do you know who I am?' he continued.

'No.'

'I am David, from the BBC.'

'Oh.'

'Do you know where you are?'

'Er, no.'

'We are in Costa Rica. We're doing a radio piece for *Breakaway*.'

'Surely not!' I responded. 'We never go further than Tewkesbury.'

Half an hour later, we had moored by a charming riverside café. The dark skins and the colourful cloths and pottery were not typical of Tewkesbury. I told David that I was going to take a birding stroll round the garden, in the hope that whatever species I saw would help me reorientate. 'Mm, House Sparrow – could be anywhere these days.' Tennessee Warbler – North America? Montezuma Oropendola – only in South America. Chestnut-mandibled Toucans! I *was* in Costa Rica!

A few weeks later, I was back in London. David called me: 'I am afraid the falling in the river thing did spoil some of the recordings. I wonder, could we sort of knock up an ending? I don't like doing this but...'

'I do!' I responded.

So it was that we sat on a bench in the Rose Garden in Regent's Park, with a washing up bowl and a flannel, with which I simulated the lapping waters of a Costa Rican beach. I was about to bid my listeners a fond farewell, when a couple of Black-headed Gulls began squabbling over a discarded sandwich and started screeching loudly. 'We can't have that!' I pontificated. 'Why not?' asked David. 'They don't get Black-headed Gulls in Costa Rica. The BBC will get letters.'

I am sure they did.

CHAPTER THIRTY

Beware of the Cat

(Dedicated to Dave)

No one has seen the Tiger, but we know he's there. There were pug marks along the main track and claw marks on a tree trunk, which he has been using as a scratching post, just like a domestic cat. Only these marks are four metres high. More than twice my height. A Tiger could reach up and sink its claws into an elephant's back. Even one as big as the alarmingly large-tusked bull that has just crashed out of the jungle 50m ahead, and now stands staring at us. Is he challenging us to approach or considering whether or not to charge? My Indian guide is clearly unnerved. In my experience, the local people are more wary of elephants than they are of Tigers. It is not surprising. Elephants don't hide. They don't skulk and there is usually more than one of them. A Tiger considers a human as food, but to an angry elephant we are enemies, or threats, especially if there are babies to be protected. It is, however, a small consolation that at least you see an elephant coming at you. You wouldn't see a Tiger. Ever. But we knew he was there.

Next morning there was more evidence. One of the lodge workers had been cycling to work when he noticed a small patch of flattened grass not far from the road. Not wisely, but probably mindful of earning a small bonus, he dismounted and swished towards the patch, shouting and flailing as he went. Tigers only function in silence, so he probably felt safe as he peered forward and saw what every safari guide and tourist is hoping for. A kill. A recent kill. Still bony, still bloody. A small deer, half-eaten, but half not. The Tiger would be back. So would we.

If I am ever asked: 'Where is the noisiest place I have ever filmed?' I would say: 'India, in town.' The quietest? India, in the countryside. Or at least, that part of Corbett National Park that morning. No rattle of distant vehicles, no rumble of planes, not even the throb of invisible airliners. A single crow of a Jungle Fowl – the ancestor of every barnyard rooster – and a neurotic yelp from a real wild Peacock, both evoked the grounds of an English stately home. The nearby trees weren't jungle, they were woods. The grassland reminded me of a deer park. Indeed, we had already seen deer, feeding nervously at the edge of the forest. They wouldn't wander out into the grassland. But we would.

We needed to find yesterday's kill. My elephant led the way, urged forward by the mahout who sat between its ears and steered by belting the animal over the head with an iron billhook. To me it looked horrendous, but to the elephant it was barely a tickle. I was perched on a wooden saddle, a foot or two lower than the highest claw scratches on that tree trunk. One cameraman stayed on the periphery, with his tripod as steady as a rock on the back of his open jeep. The second cameraman didn't have a tripod. He was shooting handheld, which should capture the reality and tension of the action. He had three angles: he could grab pictures of the animal (whatever it was), or shoot from my point of view (p.o.v.) so the camera saw what I saw, or he could shoot me and my reactions: anticipation, excitement or terror.

Slowly and inexorably we moved into the long grass, known as 'elephant grass', because that's how tall it grows. Once you are surrounded by it, it is scary. Whatever is in there you wouldn't see, unless it jumped up and waved at you. Or something worse. Against a background of silence, the only sound was the laboriously rhythmic swishing and the muffled 'flumps' of the elephant's footsteps. Now and then the mahout would apply the billhook, we would stop, and he would scan around us. Sometimes he would mutter something, presumably in Hindi. I didn't understand him, and he didn't understand me, except he recognised the word 'Tiger', which he would then repeat, with a point of a finger or a small flourish of his hand. I took this gesture to mean: 'There could be a Tiger crouching in the grass only a couple of metres away from us, and we won't see it because it is so brilliantly camouflaged. If it leaps up, it will have no trouble grabbing you and pulling you off the elephant. So let's just hope it came back last night and finished off the deer, in which case it won't be hungry any more, and is probably fast asleep somewhere in the forest.' Which was just what I was thinking! Amazing what you can convey in sign language.

For more than an hour I remained suspended between fear, anticipation and ultimately disappointment. We found the kill, and tried to convince ourselves that there were signs of recent gnawing – surely that bone was bigger yesterday? – but we couldn't be sure. We listened for the alarm bark of Spotted Deer, which is a sure sign that a Tiger is not far away. But everything was quiet. 'Almost too quiet', as they say in the movies. Another sign that the Tiger was still in the area? Maybe. Maybe not.

As the sun got hotter and the grass began to melt and shimmer with heat-haze, it was no longer Tiger time. The mahout's billhook engaged second gear and the elephant almost jogged over to the jeep, where the director was looking at the static cameraman's footage on a small screen. A wide shot of a benign landscape, bathed in gentle mist, and Bill going for a pleasant ride on an elephant. At that distance I certainly didn't look scared.

'So were you?' asked the director.

'What?'

'Scared.'

Before I answered, the soundman took off his earphones and handed them to me. I put them on. There was a noise like the gasping of a pair of industrial bellows, accompanied by a thumping drumbeat that would have filled a dance floor. 'That's you,' said the soundman. 'Breathing and heart beat.' The sound of fear.

I had never been so scared in my life. I still haven't.

LET'S FACE IT

CHAPTER THIRTY-ONE

Down on the Farm

When I was a kid my dad used to read me bedtime stories about friendly old Farmer Giles who was rosy-cheeked, straw-chewing, always cheery and above all caring and loving to not only the farm animals but also to the wildlife which he welcomed to his land. 'Good morning, Mr Badger, have a nice day.' The message in children's books was clear: farms are great for wildlife and that's how farmers wanted it.

However, as I grew into a young birdwatcher, I soon suspected that Farmer Giles had become extinct, if indeed he had ever existed. The farmers I was encountering on the outskirts of Birmingham were anything but rosy-cheeked and cheery. Some were blatantly belligerent, even physically violent. I once got clouted over the head with my own telescope while being ejected through a barbed wire fence. When I protested that I was only birdwatching, he clobbered me again. I tried explaining that his farm's sizeable pond with its miniature reedbed was something of a wildlife haven. His response was to bring in the bulldozers and fill it in.

This was not an isolated incident. Over the years, examples of some farmers' almost psychotic hatred of birdwatchers have passed into ornithological legend. Perhaps most notorious was the farmer whose fields were graced by the presence of an extremely rare bird with the pleasingly daft name of Black-winged Pratincole. It naturally attracted an ever-increasing audience of twitchers. The farmer was not welcoming. For a few days he tried to scare the bird and its admirers away by bellowing abuse in his Brian Blessed voice, followed by sporadic gunfire (presumably blanks), but the pratincole was too weary to move on – it should've been in Turkey or somewhere – while twitchers never give up unless things get really dodgy. Which they did. The farmer drove his tractor straight at them and sprayed them with what resembled – and probably was – raw sewage. A small price to pay for ticking off a Black-winged Pratincole.

For years, I considered the image of farmers to be possessive and paranoid. I admit I was often guilty of climbing fences or crawling under barbed wire, but this was what schoolboys did. What farmers did was yell at you and chuck you off. So much for Farmer Giles.

However, as I got older, I became aware of an escalating animosity that was – and continues to be – much more serious. There was clearly considerable friction and suspicion between farmers and conservationist organisations, particularly the RSPB. Probably because it had the highest profile, the largest membership and the most ideas about what was needed to conserve or create good wildlife habitat. Alas, its ideas were sometimes viewed with suspicion and construed as interfering, invasive or even threatening. They were met with the retort: 'I'm not having anyone telling me what to do with my land.' There was one farmer in Kent who every year defiantly damaged a small SSSI on his farm and every autumn spent two weeks in jail! As an isolated case, it might be amusing, but as a widespread attitude it is a disaster.

Maybe it was unavoidable. Those Farmer Giles books really didn't help. There is more to running a farm than chatting to the animals. What about the stuff about budgets and subsidies, and discussions with accountants, let alone the traumas of failed crops or diseased livestock. In a sense, farmers got lumbered with responsibilities maybe they didn't want, expect or realise. The accepted concept is that wildlife mainly lives in the countryside. Most of the countryside is farmland. Hence farmers become the custodians, whether they like it or not.

One of the most unfortunate results of the Badger cull has been the apparent taking of sides. The Government (Defra) is supported by the NFU that is 'protected' by the police. They line up against the NGOs that are largely supported by the public. This all too easily feeds a public perception of farmers as Badger killers (cullers). The truth is that some farmers support the cull, some don't. Similarly, while there certainly are farmers who are disinterested or even hostile towards wildlife and 'wildlife people', there are others who are willing to farm sympathetically, but need advice and a financial contribution. Worth their weight in gold are the farmers who know and love their wildlife, create special habitats, and even open their land for farm holidays including nature study.

It is often said that the only way of being 100 per cent certain that an area is totally wildlife-friendly is to own it. Hence the nature reserves owned and managed by such as the RSPB, the Wildlife Trusts, Wildfowl & Wetlands Trust, etc. Most of our countryside is owned by farmers. The way they manage it is absolutely crucial to the state of British wildlife. Does this government know that?

Or care?

CHAPTER THIRTY-TWO

Protecting What from Whom?

A friend of mine had just come back from a weekend in Scotland. He was excited but not totally confident, hence the question: 'Could we have seen a Golden Eagle?'

'Well, yes you *could* have done. Whereabouts in Scotland were you?'

'In the Cairngorms.'

'Well, there certainly are Golden Eagles in the Cairngorms. What was it doing?' 'We were driving. It flew across the road.'

'Are you sure it wasn't a Buzzard?'

'It was really big.'

Buzzards are big, but not as big as an eagle. Birdwatchers say that if you see an eagle but you are not sure, it is a Buzzard. But when you do see an eagle, you know! In Scotland they call Buzzards 'tourist eagles'. My friend was deflated; his wife said: 'Never mind, dear', and his daughter said: 'Told you so!'

That was about seven or eight years ago. Since then my friend has seen a lot of Buzzards, and he didn't have to go to Scotland.

He lives about 20 miles north of London and he sees Buzzards nearly every week. He doesn't mistake them for eagles, though he is getting confused by the ever-increasing number of Red Kites. It is not many years ago that both species were in trouble. In the late 1950s there were only half a dozen pairs of Red Kites and they were all deep in the Welsh Valleys. Buzzards were not as scarce as that, but they were very much birds of hills, mountains and moorlands. So much so that we assumed they needed such specialised habitat to survive.

Then, about 10 years ago, I went to the Netherlands. The polders in winter. The very epitome of flat farmland, with nary a hillock in sight. There were thousands of wild geese, and *lots* of Buzzards. Perched on posts, dawdling along dykes and most incongruously shuffling around in the grass feeding on earthworms. The conclusion was obvious: Buzzards needn't be confined to moors and mountains. The rhetorical question was inevitable: if they are happy in the Netherlands, why not in England?

And thus it has come to pass. Thanks to diminished use of poisons and pesticides, and less prejudice and persecution of raptors in general, 'tourist eagles' are proliferating. One of the secrets of maintaining what one might call their natural numbers is that they are not fussy eaters, which is just as well because they are hardly impressive hunters. They are not capable of the deadly dives of a Peregrine, or of the lightning interceptions of a Sparrowhawk, and their attempt at impersonating a Kestrel is more of a lollop than a hover. It is testament to their lack of skills that much of their 'live' food (rabbits, ducks, pigeons, etc) is already ill, or injured. Not surprisingly, Buzzards are prepared to risk their own lives by swooping on roadkill along with other scavengers such as crows, Jackdaws, Red Kites and Magpies.

So Buzzards are booming. So much so that I feared they would become victims of their own success. It is what often happens when a species becomes noticeably numerous. They will inevitably be blamed for the demise of other species, guilty or not. It happened to Magpies and Sparrowhawks – 'killing all our songbirds'. It is happening to parakeets – 'stealing all the nest holes', and now it is happening to Buzzards – accused of taking Pheasants!

Frankly, I could not believe it when the Government announced that it was sanctioning the destruction of Buzzard nests, by blasting them with a shotgun. How much blasting does it take to demolish a large platform of tightly woven twigs? And how do you make sure you avoid hitting the birds? Estate owners would also be allowed to 'remove' birds (how?) and commit them to captivity (where?).

I was not at all surprised that, within a couple of days, an enormous lobby of protest had been organised by individuals and the pertinent organisations, such as RSPB, the Wildlife Trusts and the League Against Cruel Sports.

I *was* surprised when, only a couple of days later, the Government announced that plans for the Battle of the Buzzards had been withdrawn. The fastest U-turn so far?

So what does this imply? That Defra – a department that has so much affect on our countryside and wildlife – bows to the wishes of the shooting lobby? Or that the Government really does take notice of public opinion?

Or, is it a cunning strategy? First, the Government comes up with some plan that is unthinkably absurd. Second, the public is outraged. Third, the PM makes a U-turn and drops the idea, which was so silly it would never have happened anyway. Fourth, the PM claims: 'See, we *do* listen!'

But where does that leave the wildlife? The Buzzards seem safe, for the time being. Meanwhile, the Pheasants – millions of them – are captive-bred to become living targets, to be blasted out of the sky in the name of 'sport'. Many of the birds can hardly fly and will inevitably be splattered and crushed on the road. What's left may well end up as 'Buzzard food'.

I will leave the last word to my friend: 'If they want to save Pheasants, ban traffic!' Look out for another U-turn!

CHAPTER THIRTY-THREE

Meet us, Don't Eat us

It is no secret that much of British wildlife is endangered. Fortunately, we – and our birds and mammals – are blessed by having several organisations working to conserve and protect them. Of course, this costs money, and much time and effort are inevitably spent on fundraising. However, recently I have been thinking that the wildlife benefits, but it doesn't contribute. So how's this for an idea to make wildlife pay for itself? First, you watch it, then you eat it! Let's face it; it happens already – Rabbit pie, venison, Pheasant. Wild creatures fulfilling two roles – entertainment and food. One does not detract from the other. Just because its parents are succulent doesn't mean a baby bunny is less cute, or a 'Bambi' less lovable, or even a male Pheasant less handsome. Most nature reserves have a café attached, so why not specialise in fresh local food. Very local. No transport costs, and much-needed income for conservation. Of course it needs a zappy publicity slogan. How about: 'View 'em, then chew 'em?'

I am, of course, joking. Bidding to become conservation's answer to Frankie Boyle perhaps? No, this outburst of satirical black humour was brought on by a recent visit to Iceland.

Within an hour of landing, we were taking a stroll around Reykjavík harbour. I literally could not believe what I saw. All along the quayside were kiosks and posters advertising whale-watching trips. Some of them even promised: 'If we don't see a whale, we will give you another trip – free.' There were also guarantees of sightings of porpoises, dolphins, and – arguably the northern hemisphere's most desirable bird – Puffins. Icelandic eco-tourism was obviously flourishing, which pleased me greatly since, only about 10 years ago, I had filmed, with the BBC on board, what was then the very first whale-watching boat. At the time, as one would have hoped, the commercial whaling industry seemed to be defunct, although there were rumours of its revival. But surely that was unthinkable? Whale-watching and whale butchering side by side? That must be a joke. A very sick joke.

However, less than a month ago, as I walked along Reykjavík waterfront, I realised that the 'joke' has become a reality. There were eight large ships moored side by side. The first six proudly bore the logo and name of whale-watching companies. The decks were fitted out for optimum viewing, with chairs, binocular and camera rests, and conspicuous identification charts. However, at the end of the row were two more boats, unmarked, and ominously functional. Each had a harpoon mount on the prow, a chain winch for hauling a wounded animal on board, and a big featureless grey open deck, space enough to butcher the whale. It was all too easy to picture it flooded with blood, intestines and body parts. Right now, the whaling ships were empty and silent, and some of the machinery was even a bit rusty. I even wondered if they were out of commission. Maybe the whalers had given up. Next morning, they were gone. Half an hour later, we found them again, moored even more conspicuously on the other side of the harbour. It was almost as if they were warning us: 'Oh yes, we can move, and we can kill whales. If we want to.'

The juxtaposition of whale-killing boats alongside whale-watchers was more than ironic. Iniquitous, more like. But it became almost surreal when I read the menu board outside a harbourside café. Halfway down the main dishes, it was recommending 'Whale Steak'. A rather swisher-looking restaurant was more specific – 'Minke Whale Steak'. Another one urged you to sample a sort of nautical hors d'oeuvre: a platter of 'Shag (like a Cormorant but smaller, and, I'd imagine, tougher!), Whale and Puffin'! I ask anyone who has been charmed or excited by these birds, and spent hours photographing them on the Farne Islands, or in Shetland, or indeed in Iceland, was your afterthought: 'Mmm, very cute, but I wonder what they taste like?' I think not.

But maybe I am being naïve. Maybe I should play devil's advocate. There was, of course, a time – and not that long ago – when a considerable part of people's staple diet was 'wildlife'. For thousands of years, that is all there was! Different geographical areas and different communities had their 'specialities'. Everyone knows that Iceland has always been a nation of seafarers and fishermen. Surely they have been whaling for centuries, and whale meat is a widely eaten traditional food? Well, the answer is 'no, they haven't, and no, it isn't.' Iceland only began commercial whaling in 1948, and a recent poll found that no more than 5 per cent of Icelanders regularly eat whale meat. It is certainly not considered to be a delicious 'local delicacy'. That is 'restaurant speak' to lure in the gourmets, or perhaps more likely the curious tourist rising to a dare. 'I bet you won't order blubber.' Or 'I believe in eating what the locals eat. You can't go to Iceland without eating whale, or Puffin, or Shag!' You can, actually.

I don't want to be a spoilsport, but I am afraid that seemingly frivolous attitude fuels some awful atrocities, involving willful slaughter and appalling cruelty. Order pâté de foie gras anywhere in Europe, including England, and you are condoning unspeakable torture to geese. In Cyprus, if you are offered *ambelopoulia*, just say 'no'. It looks like a jar full of pickled walnuts but its production is one of the reasons that Cypriot hunters annually slaughter literally millions of small songbirds, most of which are migrants. As indeed are many species of whale.

As it happens, Minke Whales (hunting of other species is illegal in Iceland) are thought not to be major travellers, but, like any creature that swims the oceans, they cannot be said to 'belong' to any particular country. At present, they are not endangered, but possibly only because there 'isn't much meat on them' compared to the really big ones, so they weren't considered to be worth hunting.

So are they worth hunting now? Only if there is a market for their meat, and that market is Icelandic restaurants. The demand comes largely from tourists. They have been drawn to a unique country of extraordinary landscapes, sea cliffs, bays, volcanoes, waterfalls, lakes and fresh clean air, where they can trek, ride, cycle, drive or sail, enjoying wildlife, watching birds and, of course, watching whales. If you see one surface, roll or dive, it will lift your heart. If you think of how it might die, it will break it.

There is no quick and painless way to kill a whale. It can't be dispatched with one 'good clean shot'. It will be pierced by several harpoons, dragged through the water gasping, writhing and unable to breathe. It is likely to have died through drowning before the final ignominy of being hauled out of the sea and onto that deck, soon to be awash with its blood.

As I wandered around Reykjavík, photographing menus and shop windows stuffed with literally hundreds of fluffy Puffins, I was overtaken by a couple of whales! Well, half whales actually. Both top halves. The legs belonged to young volunteers from IFAW (International Fund for Animal Welfare). They have a presence all over the world, wherever animals are in trouble. The job of this team is to intercept tourists who are returning from a whale-watching trip, explain the present situation in Iceland, ask whether or not they have tasted or intend to taste whale meat, and finally to get a signature on a 'postcard of protest', which will be added to the ever-growing pile already delivered to the government by IFAW. As the tourists return to their hotels (maybe to change their dinner order!), they are serenaded by the volunteers with a little whale song or rather 'whale chant': 'Meet us, don't eat us. Meet us, don't eat us.'

It is one of those moments that confirms my pet theory that care and kindness to animals brings out the best in humanity. Especially because the young volunteers are such a cosmopolitan bunch: French, Dutch, German, Croatian, Finnish, British, Swedish and a pair of Poles! Mind you, to be honest, everyone looks the same inside a whale costume!

My final fact-finding cruise was a day and a night on board IFAW's research vessel, a very handsome yacht called *Song of the*

Whale. She sails the world with her small crew of scientists, fully equipped with the latest technology for recording whales both at and below the surface, collecting data, and anticipating and investigating potential problems. The researchers enjoy the full trust and cooperation of the whale-watching companies, unlike the early days, when the initial response was to refer to IFAW personnel as 'terrorists', an unfortunate legacy of past anti-whaling protests that had involved a rather more buccaneering approach. No judgement, but that is not IFAW's way. On board the *Song of the Whale*, the atmosphere is anything but militant. The team are working almost non-stop, day and night, scanning with binoculars or cameras, checking video feeds and peering at computer screens. Every now and then a cry goes up, and so does a Minke Whale. It's a graceful animal. It moves with a glide rather than a leap, surfacing three or four times before sliding down to deeper waters, where the fish are. The sea is glassy calm. There is an air of concentration and contentment. This is another of those moments. The best of wildlife and the best of people.

Worth celebrating with a song. The Song of the Whale. All together now: 'Meet us, don't eat us. Meet us, don't eat us. MEET US, DON'T EAT US!'

OK, but I wouldn't mind a biscuit.

"FARMER FARMER PUT AWAY THAT DDT·NOW.
GIVE ME SPOTS ON MY APPLES
BUT LEAVE ME THE BIRDS AND THE BEES
PLEASE." Joni Mitchell. 1970

CHAPTER THIRTY-FOUR

The Way it's Gone

When it comes to environmental damage and desecration, there is nothing like the real thing. I had seen newsreels of oil spills, but it wasn't until 1993 – when I stood on my favourite beach in Shetland and watched ebony waves as viscous as lava splurging out of the *Braer* and crawling up the sand like a giant satanic amoeba – that I understood the true disaster of an oil spill.

The concept of 'acid rain' was even more amorphous until, one afternoon driving along the M4 past Port Talbot in South Wales, I saw the cause and effect. Smoky plumes of heaven-knows-what chemicals, swirling upwards from the factories towards a line of silhouetted trees on the near horizon with no more foliage on them than the deadly chimneys themselves. Black misshapen skeletons. Memorials to pollution, or posthumous entries for the Turner Prize?

I had similar revelationary shocks when I gazed upon the rotting detritus littering the north Norfolk marshes after the floods of '93;

and when I witnessed the power of the Great Storm of 1987, which interrupted my filming schedule by toppling a lime tree so enormous we couldn't get into the Tower of London!

I was mulling such thoughts, as our plane approached Kota Kinabalu, the main airport of Sabah, in Borneo. I looked down for signs of the curse that I had been told was threatening to swallow the jungle below and its wildlife with it. Palm oil. This was my first visit to Borneo, and I had not yet seen palm-oil plantations for myself. Would they show up from the air? Looking down, I noticed a few obvious brown squares where forest had been recently cleared, presumably for timber or planting, but otherwise everything looked green and leafy. Several days after landing, I realised that oil palms are green and leafy!

At first sight, a plantation looks neither unsightly nor barren. After all, they are trees, not dissimilar to the coconut palms that provide shade on many a silver-sanded tropical beach. They are stately and shapely, and their giant fronds ripple and sway like a feathery leaved ocean. You can't help wondering: 'Why aren't these as attractive to wildlife as native forest?' The answer is because a palm-oil plantation is not a mixed community of different trees. It is a congested ghetto of one species, melded together so closely to prevent light getting to the understorey, and lacking in clearings or what we'd call trunks or branches. What's more, the fruit yield is so meagre – not much bigger than a bunch of grapes on each tree – that the only way to greater productivity is to plant more and more palms, which means clear more and more forest, and lose more and more wildlife. Primary forest supports 220 species of birds here, palm-oil forest has 12!

An industry that is clearly contributing a great deal of income to the country is not going to disappear, but an awful lot of wonderful and unique Bornean wildlife will, unless the planting is controlled and the animals protected. This requires cooperation between groups of people as diverse as the forest itself. In Borneo I saw this happening.

First – in no particular order, as they say on *The X Factor* – the conservationists. The small group I was travelling with were under the aegis of the World Land Trust. Based in Britain, its role is to raise funding to literally buy wildlife-rich areas, wherever in the world they may be for sale. The Trust does not own the land. It is passed on to be managed by a local NGO. In Borneo, there is HUTAN which, unlike most NGOs, is not a mnemonic. It is the local word for 'forest'. There is the rather more elaborately entitled LEAP, which stands for Land, Empowerment, Animals, People. Which is as good a recipe of the essential ingredients for conservation I have ever heard! Oh, NGO = Non-Governmental Organisation (a euphemism for 'charity'!).

However, there is one more element whose understanding and support is absolutely crucial to the success or failure of... well, just about everything! The Government. We were granted a far from cursory audience – nay, frank and free discussion – with the Director of the Forest Department. His knowledge and concern was convincing, his confidence was reassuring and his practicality was undeniable. 'It is only because of the income we get from palm oil that we are able to finance conservation,' he said, thus complementing the words I had heard at LEAP and HUTAN: 'No point in fighting the palm-oil industry, we must work with it.' And from what I saw in Sabah, that is what is happening.

Oh, in case you are wondering about the wildlife. We saw a delightful herd of 30 Pygmy Elephants bathing at dusk (they weren't as small as I'd hoped!), and honey-furred, squishy-nosed Proboscis Monkeys staring at us as we punted by. We heard a couple of distant choruses of gibbon song, and were given a meticulous demonstration of how to build a leaf-hammock by a venerable male Orangutan, which he finished off by adding a leafy, but probably leaky, roof! Impressive primates indeed, but – during this week at least – the most impressive primate of them all was man, and woman, and children, and maybe a politician! You know, I've always said it: when it comes to good people and good deeds, there is nothing like the real thing.

SO THERE I AM, ON THE CENTRE COURT AT LAST. IT STARTS DRIZZLING THEY CLOSE THE FLIPPING ROOF! WITH ME INSIDE! SO I LAND ON KOURNIKOVA'S SHOULDER, AND SHE GIVES ME A STROKE! RESULT!

CHAPTER THIRTY-FIVE

Harris Hawks

I was recently waiting for a train at St Pancras station, when a fellow traveller approached me eagerly: 'Last week I saw a man here with an eagle on his arm,' he said. 'Are you sure it was an eagle?' I asked. 'Well, no, but it was some kind of big hawk.' He was obviously hoping I would provide a more specific identification. 'What colour was it?' I asked. 'Er, I didn't really notice the colour.' There speaks a non-birdwatcher, I thought, but didn't say! 'Was it perhaps dark grey, with chestnut shoulders, and a white tail, with a black band across the end?' He gasped with delighted bemusement, like an audience volunteer bamboozled by a mind reader. 'Yes, that was it. What was it?'

'That was a Harris Hawk,' I announced. 'It is just about the commonest bird of prey in falconry shows and zoos. They are not British, they are American, but so many of them have escaped from captivity they have recently qualified for a mention in the recent atlas of British breeding birds. If they keep escaping at this rate, they could become honorary UK citizens, like Ring-necked Parakeets.

'But why would a falconer walk his hawk at St Pancras station?' he asked. 'Almost certainly, to keep the pigeons away.' They used a Harris Hawk at last year's Wimbledon, though presumably not while they were playing. That'd put you off your serve! And the BBC used Harris Hawks to dissuade pigeons from pooing on the newly redecorated Broadcasting House. Not everyone approved, but a BBC spokesman promised that no pigeons would be harmed. Unfortunately, one hungry hawk couldn't resist pouncing on a particularly plump pigeon, flying off to the roof with it, and tearing it to pieces, before the horrified eyes of a number of BBC staff. What's more, the supposedly well-trained hawk refused to fly back to its handler. Of course, when it finally did so, it was fired. Actually, there is a fortuitous irony in that these days there are several pairs of wild Peregrine Falcons breeding in central London who dine almost exclusively on Feral Pigeons, and whose territory almost certainly includes several railway stations and Broadcasting House. So that should put an end to the scandalous business of overpaid American raptors coming over here to eat our birds, which, as it happens, are probably not to their taste anyway.

The fact is that Harris Hawks are not the deadly feathered Exocet missiles that Peregrines are. Indeed, back home they hunt at a rather leisurely pace, in cooperative groups. I don't mean feeding flocks, like some of the small falcons catching dragonflies and such. Harris Hawks go in for properly coordinated teamwork. Hunting parties are between two and six birds. One of them flies ahead to look for potential prey, more likely a Jackrabbit than a pigeon. Then a second bird moves up and takes over the searching. Once a victim has been spotted, they overtake each other until they have surrounded their supper, which understandably freezes with fear and – acknowledging that all escape routes are barred – surrenders to its fate.

I have seen wild Harris Hawks in the Arizona desert, swooping and plunging against a background of a burning blue sky, sagebrush and giant cacti. It struck me more as an aerial ballet than a hunting party. The local Jackrabbits, however, no doubt would not agree. Any more than a prehistoric lizard would welcome the attention of a team of velociraptors. At which point you may fairly ask: 'What on Earth have Harris Hawks got to do with dinosaurs?' I shall tell you.

Several years ago, I presented a short series called *The Truth About Killer Dinosaurs*. It inevitably featured velociraptors. In the film *Jurassic Park* they were depicted as quite big, which they weren't – they were about the size of a turkey – and rampantly vicious, which they may well have been. One thing was undoubtedly true: they hunted in packs, pursuing and surrounding prey, before pouncing, killing and eating it, very much in the manner of… Harris Hawks. It is, of course, now widely accepted that birds evolved from small dinosaurs. So there we are, from *Jurassic Park* to St Pancras station is not as long a journey as you might think! Pity it only goes one way.

CHAPTER THIRTY-SIX

First Mornings

Show-business people talk about first nights. Birdwatchers talk about first mornings. Here is a classic scenario. You have gone abroad for a holiday to somewhere new. You arrived at the hotel, lodge or resort in the evening when it was dark. This may be a little frustrating, but it is also tantalising. What is it going to be like out there when you can see and hear?

Over the years, I have had a few surprises when the sun came up! On my first trip to India we were escorted down a pitch-dark path into a concrete cabin and fell asleep with nary a clue what kind of terrain we were in. I staggered out at dawn and almost fell into a cesspit. The considerable consolation was that there were several Painted Snipe feeding on it, and, even better, the surrounding woodland was pleasingly open and full of birds, many of them species that occur in Britain but rarely (Bluethroats, Red-breasted Flycatchers, etc). They were easy to hear and easy to see.

This is rarely the case in rainforest, wherever it is. Heaven knows, I am not dissing its importance or the profusion of its biodiversity, but the fact is my worst ever first morning was in Papua New Guinea, when I tramped the jungle trails for three hours without seeing or hearing a single bird! My wife Laura accompanied me on that trip. As, indeed, she did to Kenya, where we shared a memorable first morning at Lake Baringo Lodge. I had been aware of occasional snufflings and grumping in the night, but had not expected to draw the curtains and find a hippo with its nose pressed against the window. At least I could say: 'See, it wasn't me snoring.'

Challenging though it may be, there are few experiences more daunting for a birdwatcher than emerging on the first morning and not being able to recognise a single song or call. Of course *seeing* the bird should help, but not necessarily, especially in Asia or South America where there are whole families that don't seem to bear any relationship to anything we have at home. Page 96 of *Birds of Brazil* includes reedhaunters, foliage-gleaners, Firewood-gatherer and barbtails. And guess what? They are all little brown birds.

So what was my best first morning ever? Rather bizarrely, it was at a 'Dude' Ranch in Arizona. We had arrived in darkness, made even more disorientating by clouds of dust kicked up by every vehicle. The Ranch 'Cookhouse' was like the buffet at a Holiday Inn, and our chalet – far from being rustic – was a stone edifice that strangely resembled a Henry Moore sculpture. None of us was happy. The cost of the accommodation was in return for my wife writing a flattering travel piece – so that could be awkward – and my daughter Rosie kept reminding us that she was the only 10-year-old girl on Earth who hated horses. I myself was suffering a period of low mood and not sleeping well.

However, my insomnia was a blessing in disguise. As soon as I sensed a glimmer of light, I slipped outside into what was in effect a giant rock garden. A small-scale version of the true Arizona desert where cowboys roam, the Coyotes howl, and the cacti reach for the sky. I intended to visit the real thing, but for the moment this would do nicely. The birds clearly agreed. There were Cactus Wrens everywhere, same shape as ours but huge and stripy. There were woodpeckers – cactus peckers more like – laughing as woodpeckers do, and popping and peeping in and out of their prickly nest holes. A parade of Gambel's Quails danced across the path sporting bouncy little quiffs on their heads. All around me there were warblers, flycatchers, orioles, sparrows, shrikes and raptors. All different from home, but just that little bit the same. I think that was why I enjoyed that walk so much. As I ambled down the path, the species revealed themselves one by one. It wasn't overwhelming. I had time to take notes and do sketches. I rifled through the field guide over breakfast, and announced that I had seen more than 20 'lifers', including my first Roadrunner.

The rest of the week was not so good. It was wet, cold and almost birdless. Laura lied ('fab for families') and Rosie fell off a horse.

CHAPTER THIRTY-SEVEN

Going Bananas

I'm no stranger to dressing up as a mammal or a bird. My first experience was at primary school when I was cast as a baby dragon in the school play. I do not remember the play, but I do remember the costume. It was like a scaly-patterned 'onesie'. It was tail-less, and I was therefore puzzled when our comely drama teacher asked me to spice up my exit with a vigorous wag. I protested that I had nothing to wag, but she insisted that I had. To this day, I am not sure if I was being dim or provocative. All I do know is that she finally had to resort to using the 'b' word, at which she blushed and the class fell about with joy. 'Miss! You said "bum"! Teehee.'

I was clearly traumatised by the experience, as I went through secondary school and three years at university without feeling any compulsion whatsoever to wear scales, fur or feathers. However, I more than made up for that during the 1970s, The Goodies Years as I like to think of them, though I am not of course insinuating that nothing else happened during the seventies. Actually, it was a decade in which wearing animal costumes was all too rife – reprehensibly, in the case of supermodels, and gleefully in the case of The Wombles. In the case of The Goodies, it was downright peculiar.

In the episode with the giant kitten, all three of us spent several days dressed as mice. I also gave solo performances as a Dalmatian dog, and as a Dodo (I did of course become extinct), and Tim and Graeme became intimate inside a pantomime horse, a role which I enjoyed greatly, since I got to ride them and give them a damn good thrashing with my whip. Comedy is rarely painless, but if I had to choose the least enjoyable animal costume I have ever had to wear it was when all three of us played giant rabbits and re-enacted extracts from *Watership Down*. Being run over by a lorry was pant-threateningly scary, but at least it was soon over (after seven takes!), unlike the seemingly never-ending, ever-worsening torture of being slowly broiled in our own sweat, leaving us so soggy and smelly that even our loved ones were repulsed.

So, if dressing up as a rabbit is uncomfortable, what about a gorilla? There must be plenty of opinions because – let's face it – unlike the magnificent animal itself, people in gorilla suits are not exactly an endangered species. One of the reasons is that costume hire companies don't know their primates. You can ask for anything from a Chimpanzee to a gibbon, but they'll always send you a gorilla. (You can imagine my pedantic strops when promoting 'The Funky Gibbon'.) You will see gorillas at Halloween, gorillas in panto, gorillas in tutus and gorillas in top hat and tails (which is ironic, 'cos they don't have tails. Any more than I did).

I don't think many people would disagree with the statement that as far as gorillas are concerned you can't have too many of them. It is a fair rule of comedy that 'more is funnier'. One dwarf is unremarkable, but seven are a hoot. One pantomime horse may get a smile, but a Grand National of them brings the house down. One gorilla is probably less than hilarious – no matter what it is wearing – but a whole rock band of five or six gorillas will keep an audience whooping for several songs before the truth becomes obvious – that it is almost impossible to play a guitar wearing a gorilla suit! More gorillas than half a dozen and the rule kicks in: the more the merrier!

The greatest gorilla gathering on Earth convenes in the region of Tower Bridge in London on a Saturday morning in late September, when there are more men and women in gorilla suits than there are endangered gorillas in the wild. The event is known as the Great Gorilla Run and it is indeed great. The laughter, good humour and waving of bananas at the start is truly uplifting. The state of the runners after an hour or two is something to be pitied. The effects of hopping in a rabbit suit are positively cosmetic compared with running as a gorilla. Rabbit-fur-induced perspiration is as fragrant as fine cologne compared with the fetid stench of gorilla sweat. Added to which, if the runner isn't instantly released from his or her costume, there is serious danger of being suffocated or drowned. One of the few consolations is that hair becomes so drenched that it can mimic the effect of gel or Brylcreem, which can look slick on gentlemen. Ladies should not wear make-up to avoid emerging like one of the Addams Family, unless you fancy going Goth.

By the way, the reason I know such details about the annual Great Gorilla Run is that I am a regular attendant. Indeed, I wouldn't miss it for the world. Nor of course would I dream of actually taking part. Such exertion would almost certainly prove terminal, and then I wouldn't be able to attend the following year, to send them off at the start and – even more fun – allow them to collapse on me at the finish, where I also adorn them with their well-earned gold medals. No silver, no bronze; they are all winners to me! That's what I tell them every year. Those who are still able to utter any coherent sound – apart from gasping, retching or weeping – usually greet that line with a howl of derision, signifying that, despite nearly killing themselves, their sense of humour is alive and well. The returning runners are molested by news cameras and reporters searching for statements of the obvious: 'Why do you do it?' 'Why is it important to save gorillas?' 'Aren't you tired?' In reply, the clichés also abound: 'Tired but happy'. 'No pain no gain'. 'Well, it's for a good cause isn't it?' I heartily agreed with the young lady who was both draped and dripping all over me, which wasn't surprising since she was dressed up as an elephant. As she flirtatiously nuzzled the tip of her trunk in my left ear, I was tempted to quote my old Python chums: 'This is getting silly!' Indeed, and all the better for it.

Alas, it is also getting serious for the gorillas. Please help.

CHAPTER THIRTY-EIGHT

Crimes Against Nature

How do you kill a bird of prey? Shoot it. Trap it in a cage, then shoot it. Poison it, by putting out bait laced with a deadly chemical. Or wipe out a potential family by destroying the eggs or stealing them, and selling them to egg collectors or to foreign breeders who will hatch them and rear the birds for falconry. All these activities are – and have long been – illegal. All have been going on for many years. All are still rife.

In the majority of cases the crime is committed by a gamekeeper. His job is to protect the 'game' birds that are also, of course, destined to die, bagged by 'sportsmen' who will pay handsomely for the privilege. Not even the most myopically sentimental nature lover could deny that the diet of some birds of prey does include gamebirds and their chicks and even eggs. Predators are of course part of the 'natural balance' of things, but on a shooting estate the aim is to maintain an unnaturally large population of – literally – 'target' species. Therefore, we could hardly contradict a gamekeeper who argues that he is … just doing his job.

But there could be some aspects of the way he is doing the job that would make him a criminal, and if caught he would be liable to a fine, or possibly even a prison sentence. He may or may not lose his job, but if he does it would be iniquitous indeed, since whoever fires him, probably hired him and instructed him to 'control' – i.e. kill – raptors. He could certainly plead … 'I was only carrying out orders.'

But he is still guilty as charged, and criminal justice is often rough. The street pushers are imprisoned, while the drug barons stay free. The 'hit men' take the rap, while the 'Godfather' is immune. On some estates, the gamekeeper gets caught, while the landowner feigns innocence or indignation. The truth is that it is he who should be facing prosecution.

It is generally agreed that the most productive wildlife locations are those where the land literally belongs to an owner with a genuine environmental ethic. It is why RSPB appeals are so often concerned with land purchase. Fortunately, there are many private landowners who are knowledgeable, conscientious, enterprising and generous. Unfortunately, there are still… *a lot who are not.* Some of them even regard the RSPB and other environmental NGOs as a sort of traditional enemy, as if they are engaged in some kind of deep-rooted sectarian war. It is the same mind-set that regards any bird with a hooked bill as 'vermin.' It is on or near those estates that you will hear the gunshots, or see the poison bait or snares, and witness the corpses, not just of birds of prey, but any other bird or mammal that might conceivably take a gamebird.

Many wildlife protection laws exist. Others, however, have been mooted, and yet have mysteriously stalled. The intention to criminalise possession of various poisons was announced five years ago, but as yet the names have not been specified. Why not? Everyone involved knows what they are. The police and the NGOs are doing a terrific – and often dangerous – job. There is widespread public approbation and cooperation. Offenders are being caught and punished. And yet the cruelty continues. Why? Because people with money are offering money and making money. Gamekeepers know they are not above the law, but there are landowners who may think they are. They must be named and shamed. A tabloid cliché perhaps, but it is what I believe should happen, even if the social status of some of the names may cause controversy and embarrassment.

When I first started birdwatching more than 50 years ago, many of Britain's birds of prey were very scarce. Red Kites were down to a few pairs, restricted to Wales. There were even fewer Marsh Harriers, only in East Anglia. Peregrine Falcons could only be found on craggy sea cliffs and were not common. Neither were Golden Eagles. And Hobbies and even Buzzards were a rare sight. There was a single pair of Ospreys at Loch Garten, and White-tailed Eagles were unknown as a breeding bird, although I did see one sitting in a field in Norfolk in 1958. (The eagle was sitting in the field, not me.)

The scarcity of these species had been caused by widespread DDT poisoning, though it was not specifically aimed at them but was an ingredient in various pesticides used by farmers. No doubt there was also some intentional poisoning, and certainly much gamekeeping 'control'. Fortunately, outrage, research and legislation began to get the situation under control and slow down the declines. RSPB schemes, creating habitats, pioneering reintroductions and tirelessly pursuing criminal elements, not only increased the numbers of birds, but also seemed to encourage some species to immigrate in larger numbers. Ospreys are now almost as ubiquitous in Britain as they are around the world. Marsh Harriers are now to be found building nests in tiny patches of reeds or crops, illustrating the principle that when a species increases its numbers it can't afford to be so fussy about its habitat. Peregrines have moved into cities. Of course, the once extremely rare species that most people have now seen, even if it's only swooping over a motorway, is the Red Kite. In parts of Scotland you are probably more likely to see a White-tailed Eagle than a Golden Eagle, or possibly both species circling on a thermal.

Alas, though, you are almost equally likely to find one dead. The battle continues.

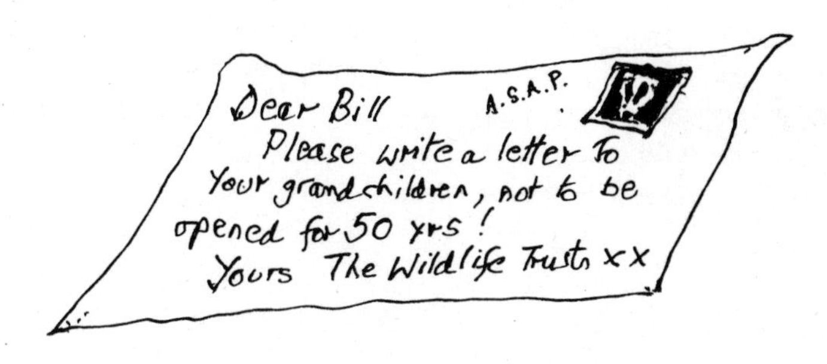

CHAPTER THIRTY-NINE

Talking About your Generation

Letter to my grandchildren, as requested by the Wildlife Trusts on their 100th anniversary (not to be opened for 50 years, unless you can't wait).

My dearest grandchildren

I am writing this in 2012. I am 70. Not as old as the Wildlife Trusts, but old enough to have seen a lot of changes. Throughout my teenage years and beyond, I was an obsessive birdwatcher. If you had asked me then: 'What is the state of the British countryside and wildlife?' I would have said: 'It seems OK to me.' I saw lots of birds on farmland, flowers in the meadows, frogs and fish in the streams, marine life in the rock pools, seabirds on the cliffs and so much more, and it all seemed fine. Looking back now, I was surely dwelling in blissful ignorance. I wasn't the only one. The principal agenda of a naturalists' club was to enjoy fun and fascination, not to fret over problems and dilemmas.

Then, in the 1970s and 1980s, I – and everyone else – had to learn a whole new vocabulary. 'Nature study' became 'environmental studies'. Surveys and analyses begat figures and facts, which spawned such ominous words as 'declining', 'threatened', 'endangered', and – gloomiest of all – 'extinct'. The grim truth was emerging that there was nary a species nor habitat that was not cause for concern. Knowing this, it was inevitable that for many naturalists their hobby became a crusade.

Now kids, at this point you are probably expecting me to apologise on behalf of my generation, but I am not going to. The state of the wild nation wasn't our fault, except in so far as we might have been a bit slow to notice. But that is often how it is with change. Like Joni Mitchell sang: 'You don't know what you've got till it's gone.' She was singing about woodland ('they've taken all the trees and put them in a tree museum') but the same could apply to heathland, downland, wetland, or any kind of land, or to any of the creatures that live there. At least we realised what we'd got. Maybe just in time.

People of my age are apt to comfort themselves and try to impress young folk by revelling in the (almost) unarguable achievements of the 1960s and 1970s. Ah, to be a teenager with The Beatles and The Stones, to witness the birth of satire, to be buying records during the decade of the great seminal LPs, and to be a viewer during what will likely be judged as the golden age of television. Frankly, kids, we never had it so good – and you certainly haven't – and I am proud to have been part of it. But we old folk have one more claim to fame. Or is 'fame' too ephemeral, too frivolous? 'Satisfaction' is a nobler sensation. Yes, the sixties and seventies gave us magical music and terrific telly, but it also saw the proliferation of environmental awareness. Knowledge.

No longer could polluters, poisoners or desecrators claim that they didn't realise that what they were doing was wrong, or that they didn't know how it could be put right. It continues to be true today. Even on a massive global scale, we are aware of both the problems and the solutions. Whether knowledge leads to action depends on the resources – the money! – the will and the passion, and above all the power.

Now, kids – and your kids if you have any (two maximum please) – I don't know if you care about the natural world, but if you do and you want to actually get things done rather than just have an opinion, then please never forget those two words: 'passion' and 'power'. The power can refer to 'people power', public opinion and protest. Never let it be underestimated, but its limitation is that while we punters may influence the lawmakers, we do not draft, pass or enforce the laws. You – and I – have the passion, but not the power. What the world needs are more people with both. Genuinely, honestly. Not a pretence, the real thing.

Do you get my drift? I am trying to give you a little career guidance. Not just to you kids, but to anyone. Assuming you love and care about wild things and wild places, what kind of job should you aim for? A warden of a nature reserve? No. In the PR Department of an NGO? No. A farmer? Mm, better. Prime Minister? Yes! Minister for the Environment. Now you're talking. The minister of anything? Any MP? Yes. A housing developer? A town planner? CEO of a multinational oil company? Richard Branson? Rupert Murdoch? Simon Cowell?

Or am I just being silly? Well, all that lot have the power. If any one – or better still three or four – of them had a passion for wildlife, I think it would help. Don't you? Oh, I don't suppose you remember any of them! Lucky!

Keep it natural.

Love,

The Green Grandad

PS Did they ever build HS2?
PPS Did they ever reintroduce Wolves?
PPS Did they ever have to cull the kites?

GRANDAD BIRDWATCHING IN HIS YOUNGER DAYS.

FALLING OFF A BLOG

TWEETS PODCASTS FACE BOOK
BLOGS INSTAGRAM YOUTUBE
BUT I HAVEN'T GOT ANYTHING TO SAY
THAT DOESN'T STOP EVERYBODY ELSE!

This is a message to anyone who used to follow my blog on my website – I am sorry that I gave up doing it, but at least I hope the rest of this book brings back pleasant memories, so you won't mind reading it again. If you didn't enjoy these 'posts' the first time, I can only plead that you give them a second chance. If you are not aware that I ever had a blog, that is good, because this means that these pieces will be entirely new to you. I hope you enjoy them.

By the way, the reason – or excuse – for giving up blogging was not because of laziness, forgetfulness or lack of interesting incident in my life. It was simply that it got too much for me once I started writing a monthly page for *BBC Wildlife*, providing quotes for NGOs' press releases, tweeting about everything from badger culls to Manchester United, and discovering that fans of Reading Football Club sing 'Bill Oddie, Bill Oddie, rub your beard all over my body!' When I heard this, I was flattered yet fearful. I warned them it could be a curse. It was. They were relegated. But they will be back. And some day, so will I.

Meanwhile, here is a selection of what I called News of the Wild. Note the often slightly salacious headings to each paragraph. They were intended as a tribute to the Rupert Murdoch school of journalism.

BLOG ONE

News of the Wild

Unnatural urges

I have a small dilemma (ooh, missus!). The title of this blog, 'News of the Wild', is, of course, a deft and ironic reference to a recently defunct Sunday newspaper famed for featuring smut and scandal undiluted by accuracy or truth. It rarely covered wildlife or the environment; hence the irony. However, as I hinted in last month's blog (still available on archive) I am developing a tendency to become more diversified myself. To put it bluntly, less wild and more news, whatever the topic. I confess I feel a bit as if I am rejecting the religious in favour of the secular, but it's a decision brought about by two things: first, I now have my very own page each month in the unarguably authoritative and sumptuously illustrated magazine *BBC Wildlife*, which satisfies my 'natural' urges, as it were. Second, I feel justified in making this and future blogs as random, as frivolous or as contentious as I like. Hence, my dilemma is should I now call it 'News of the World'?! Am I allowed to? Or would I be in trouble with Rupert Murdoch? Actually, he might well be grateful to me for taking the name off his hands. But then I'd probably get investigated for phone tapping. I think I'll stick with 'News of the Wild'. Apropos of which...

Dig

I have been conducting a little Grey Squirrel experiment. Don't worry, nothing unpleasant. We all know that Grey Squirrels (and Red ones) gather up acorns at this time of the year and then bury them all over the place – especially in my lawn – so that if the winter gets harsh and food gets short, they can dig up their hidden stash. I am sure all of us have wondered whether or not they are clever enough to remember exactly where the acorns are buried, or do they just scamper around digging more holes until they get lucky? The answer is 'no' and 'yes.' No, they don't remember and, yes, it's just potluck if they find any acorns, which is why they bury so many.

Lovely acorns

Nevertheless, I have to agree it is a pretty enterprising strategy that benefits many hungry squirrels and plants a few new oak trees at the same time. Which brings me to my experiment. The nearest oak tree to my house is about half a mile away, and to get to and from it entails crossing two or three busy roads, whether you are a human or a rodent. I decided to make life easier for my garden squirrels. Thus, I returned from my morning saunter on Hampstead Heath having collected a bulging pocketful of acorns. I went out into the back garden – causing two squirrels to panic and fall off the bird feeders – and I placed the acorns carefully and conspicuously on a dilapidated birdtable that I have not thrown away in case it comes in handy for scientific experiment, like this one. I uttered a lilting cry, such as was once used by cockney street vendors: 'Get your lovely acorns. Sweet and succulent, straight from the oak tree!'

Tails atwitching

Even as I retreated indoors, there were two curious squirrels peering over the garden fence with tails atwitching. I peeped out through the back window. Squirrels on the fence. Acorns on the table. How long would it be before they descended on their personally delivered breakfast? I waited half an hour. Squirrels on bird feeders. Acorns still on table. I went up to my office and did a couple of hours' writing. I came down again and looked out of the back window. Squirrels gone. Acorns hadn't. I went outside and counted them: 35. One more than I put out! I recounted: 34. OK, the same as I put out. Later in the day, there were still 34. As dusk fell, the same. Next morning: three squirrels now, but still 34 acorns. Two days later, and the pile appeared still intact, though my count revealed that two acorns were missing. I found them on the floor.

Repugnant

This morning, after four days, about half of the acorns have gone, been disturbed, or in some cases slightly nibbled. I suspect Jays. So what do I deduce from this experiment? That Grey Squirrels don't like acorns at all? Indeed, the very aroma of them is so repugnant that they bury them rather than tolerate it. Only as a very last resort, in winters when starvation is nigh, will they dig them up and endure the ghastly taste of emergency rations.

Or is it that Grey Squirrels prefer good-quality bird food? I could help them there!

Marketing ploy?

Talking of bird food, I was in my local garden centre the other day when I happened to accidentally peruse the range of bird foods produced by a company whose brand name shall remain unrevealed, but wasn't Haith's or me. What caught my eye was the fact that nearly every one of a large number of bags bore a picture of a different species and that the claim on the bag was that the food therein would attract the species depicted. There was a specific food for Robins, Blackbirds, Song Thrushes, Blue Tits, Great Tits, Goldfinches, 'finches', swans, ducks (what happened to geese?) and sparrows (meant to be House Sparrows but the picture was of a Dunnock). Now, I am not insinuating that this is some kind of marketing ploy, and certainly not that the product is below par, but let me assure you that you don't have to purchase a dozen or more different foods to attract a good selection of birds. There are certain groups: ground feeders, seed-eaters, soft-food eaters, but beyond that, put it this way: Blackbirds, Song Thrushes and Robins have much the same tastes. As do Greenfinches and Chaffinches, and so on.

Of course, the Grey Squirrels will eat anything that isn't meant for them!

Done it again

Well, I've done it again! I allude to a blog full of laughter and lasciviousness and end up writing about garden wildlife. Next time, next time; I promise.

Are you sitting comfortably?

But before I migrate south for a few days to the Isles of Scilly… The other morning I was having brekky in our local tea-room. At a nearby table, a Hampstead mum was reading a ToyTown book to her toddler-aged son. He wasn't toddling then, he was sitting and listening intently, and I admit so was I. Imagine my surprise when mum showed a picture to the boy and said: 'Look, Noddy's got a satnav!' Unfortunately, mother and child departed, plus the book, and I hadn't the nerve to call after her: 'Noddy's got a satnav! Is that true?'

I hurried home, not certain which was shocking me most, that Noddy's car had had a 21st-century makeover, or that I'd started hearing voices about updating children's stories. As luck would have it, my wife, Laura, writes for children's television. She would not only know about such modernisations, but may even be responsible for some of them.

Disaster

It probably wasn't the first question she expected to hear when I got home: 'Laura, does Noddy have a satnav?' She had to think about it. Then she made a pronouncement that I felt had just a tinge of defensiveness about it, almost as if I had insulted the integrity of her profession. 'I am sure Noddy doesn't have a satnav. We don't change traditional characters. Of course, Bob the Builder has a mobile. And in one series he did have a laptop, but that was a disaster.'

I bet it was. And as for Winnie-the-Pooh getting hooked on World of Warcraft… And have you seen Angelina Ballerina's Facebook page?

BLOG TWO

Frogs' Porn

Who's been eating my frogspawn?

I know it is only a small pleasure, but a few weeks ago I was so happy when one morning I pottered into my garden and discovered that the smallest and shallowest of my five ponds contained not one but two tightly clustered dollops of frogspawn. I was not only delighted but also surprised, since I was yet to see a frog in the garden this spring. However, a few minutes later, a telltale plop and ripple revealed there were in fact two frogs in the slightly larger pond about half a metre away. I must have overlooked them, which is understandable since that pond is not only larger but also much deeper. By the way, if you'd prefer more precise dimensions, the spawn pond is about a metre long and a couple of inches deep, while the larger one is maybe a centimetre more in length, but is a veritable garden Loch Ness, with a depth of at least a foot. While I'm at it – for the completists among you – my third pond is perched in the rockery like a diminutive mountain lake, and, like many real mountain lakes, is devoid of wildlife. Number four is ready-made, moulded and rather dull. It is adjacent to the bird feeders and is consequently used mainly for drinking and bathing. By the birds, that is. The waters are continually agitated by a small fountain, which sprays out unpredictably and randomly, but intentionally as I have amended the 'spout' by fixing a small piece of hollow wood over it, which is riddled with holes and therefore expels water all over the place, especially when the whole fragile arrangement is knocked skew-whiff by a clumsy pigeon (is there any other kind?).

The big pond

Finally there is the 'big pond' at the end of the garden. This used to be the one I dutifully dug out when we first moved into the house about 30 years ago. The previous owners had built a raised concrete patio in about the daftest available site which – due to a canopy of vast trees in surrounding gardens – received no sun whatsoever. It would also have been a pretty lousy place for a flowerbed. One might have thought that the same shivering shadows that prevented anything blooming or blossoming would also have deterred anyone wishing to bask, read or snooze in a deckchair. To be honest, when I first saw the supposed patio it didn't look as if anything living had visited it for years, except the snails, centipedes and woodlice that lurked or scuttled under the pile of old plant pots.

But I knew what I was going to do with it. I had never had a garden pond. In fact, I'd never really had what you might call a garden. Now that I had, I was overwhelmed by a pioneering spirit. Anyone who can remember the day they dug their own pond will no doubt recall the many emotions. The exhilaration, the anticipation, the stress, the throbbingly aching limbs from the digging, and the ever-worsening feelings of exhaustion and nausea that come from being hunched over for however long it takes to create a hole long, wide and deep enough. In my view, it should take a day. It might take less, but it must not take more. It is a matter of pride to be able to announce: 'I've built a garden pond!', whereupon you will be asked: 'How long did it take you?' To which the imperative answer is a nonchalant: 'Oh about a day. Well, less than a day actually. About a morning.'

Your audience will be impressed and doubly so when you amend the time scale to: 'Well less than a day actually, not much more than a morning.' Hopefully this will encourage, challenge or indeed oblige them to have a go in their garden. If they do that's great, but if they don't, it is your next duty to make them feel guilty. 'It'll take less than a morning, honestly. Garden ponds are really important. Not just for frogs. All sorts of things.'

Hosepipes ahoy!

My big pond was a major feat of engineering, involving the wielding of two sledgehammers, the gouging out of several weighty bucketfuls of London clay, and the laying of a butyl liner (butyl is rubbery waterproof stuff), and the final filling up process (no hosepipe bans in those days) which was carried out at least three times, until I finally got the water level level, if you see what I mean. It was a lot of work, requiring a degree of expertise, which I didn't have, but it was indeed basically completed within one day, albeit the final filling had to be conducted after dark by the glow of a couple of garden lamps. I also left the back door open so that music from my hi-fi drifted out which, coupled with the burbling hosepipe and the flashing lighting (I fixed the dodgy wiring later in the week), combined to create a passably impressive 'son et lumière'.

There is, alas, a slightly ironical postscript to the story of my big pond.

It is still there, but not in its original form. The butyl lining lasted for a couple of decades until it started to leak and the water began to seep away into the clay. I immediately resolved to repair or replace. However, after 20 odd years, the butyl lining wasn't the only thing that had deteriorated. So had I. My sledgehammer and digging days were over. I had by then created my several little ponds. I'd dug one, but for the others I had used those moulded plastic substitutes you can buy at the garden centre. Once put in place (I recommend a spirit level) and furnished with a few aquatics, they rapidly attracted a modicum of wildlife, most notably, of course, the frogs. So the question now was: how big a moulded pond can I get? The answer was 'flipping enormous'. To get one home in a car, to quote *Jaws*: I was gonna need a bigger car! Or a smaller pond. Hence the size of my new big pond was dictated by what I could get into my Renault Clio.

Prefers full sunshine

It was a bit sad placing a plastic pond in the hole left by what had been a 'natural' one, but by the time I had disguised the edges with rocks and branches and put a few little pots of aquatics on the ledges, it looked reasonably authentic. A slightly more powerful fountain than I had on the 'bird-feeder pond' added a classic running water sound, just like on the meditation CDs, while the permanent cascade almost qualified it as a 'water feature', like Alan Titchmarsh talks about. But of course this wasn't a 'water feature', it was a 'wildlife pond'. A title it totally merits, except for one thing. At no stage during its first two decades as a natural pond, nor during the several years since it went plastic, has my big pond attracted anything that could be judged to be wildlife! There are probably a few woodlice under the damper logs, but I have never ever seen anything truly aquatic, either in or on the water. There is, of course, an obvious reason. Just as a corner that never gets the sun is a pretty useless place to put a patio, or a flower bed, it is also a pretty unproductive site for a pond!

Dollops!

On the other hand, a small shallow pond, created by sacrificing a little patch of lawn that is in the sun almost all day, is 'frog paradise' and 'spawn city'. The double dollop was only the start. Next morning, there were two or three more portions; the next day the quantity had doubled, even if the frog count hadn't. Most springs there is one, usually rainy and mild, night, when it seems that all the frogs in a particular region feel the primal urges rising in their froggy loins and set off to their ancestral mating ground, or pond, where they indulge in what can only be called an orgy. It is not romantic, and it doesn't even look enjoyable, as males grab and grapple with females, until some of them are literally suffocated or drowned. One is tempted to comment: 'I've heard of a bit of rough but this is ridiculous!' Slightly disturbing though it be, I rejoice at the yearly gang-bang when my littlest pond is one day a seething cauldron of amphibian lust, and the next morning a very large helping of tapioca. This year, however, it wasn't like that.

The spawn bank

Or rather, there was only half of the equation. I never did see more than two or three frogs, and the only coupling was a soft and sensual demonstration of the request: 'be gentle with me'. And yet every day the spawn bank got bigger and bigger, until the pond was quite literally full! At this point, the frogs decided their work was at an end – as well they might – and I have not seen them, or any others since.

The circle of life

What I did witness was the beginning of one of the few life cycles that, as they say, every schoolboy (and girl) knows, and has probably witnessed. It wasn't long before several of the spawn clusters began the process, as the little black dots got a bit less little, and acquired a smidgeon of a tail shape. One particularly sunny morning stimulated the first wrigglers, and gave me a hint of what a tureen of life and energy my pond would eventually become, not to mention the hopping hordes of froglets that would invade my lawn.

It was going so well

Then something puzzling happened. They all disappeared. Dots, tiny tadpoles and spawn. The lot. All gone. No, I exaggerate. In fact, after swishing my hand gently in the weeds and water I disturbed one single tiny tadpole. I caught it in a cup and plopped it into a small tank along with a dozen others that I was rearing as exhibits to entertain my grandchildren. And myself. As I did so, it struck me that the ones I had rescued were still alive and well. Whatever had happened to the others, hadn't happened to them. But what had happened? I rummaged on the internet, but basically only confirmed what I had figured out for myself. Was it infertility, or was it what one might call lack of manpower? Male frogs don't impregnate the female and leave her to lay eggs, job done. She produces the spawn, and he then has to do the rounds, fertilising as much of it as he can. I sympathised with the male in my garden. There was a veritable spawn mountain, and possibly only him to fertilise it. Maybe he ran out of juice, as it were, or energy. I would not have blamed him, but the evidence didn't fit the theory.

I had noticed that some of the dots in the eggs had turned milky white and were presumably dead. That was most likely the result of a couple of days of late cold weather, which even included snow. In addition, some of the 'jelly' was limp and lifeless and could have been infertile, or rather unfertilised. But by no means all. Out of a pond full of spawn, and thousands of potential tadpoles, some casualties are to be expected, but there was no sign of any escalating epidemic or impending massacre. The depletion I had seen so far seemed perfectly natural. Until that morning, when *everything* disappeared. Except my little sample colony. They were fine in their tank. Whatever had got everything else hadn't got them. Surely that meant it was neither freeze nor disease.

An acquired taste

Was it possible that the whole pondful had been eaten? Or should that be drunk? To be honest I have never really thought of tadpoles as food or prey, but think of them as aquatic worms or slugs, or indeed small frogs, and they must surely get gobbled up by, for example, birds. I had noticed a queue of the usual suspects lurking and lunging by or even in the pond. Magpies and Jays were circumstantially guilty, but the only species I actually saw grab a 'taddy' and devour it was a Robin! No doubt quite a number slid down into that plump little scarlet tummy, but surely not all of them! And what about the glutinous globules? Imagine pecking at a pile of transparent jelly. It would take birds a week to get through it. Or had there been an enemy within, under the water? I don't think so. I keep no fish and I have never seen a newt in the garden.

So whodunnit?

A possible clue awaited me down by the shed. I confess I have a bit of an aversion to throwing away household chattels and furniture. This isn't for environmental reasons. I could, of course, take it to the recycling centre. However, I would rather incorporate suitable items into the decor of the garden, as it were. Thus, there are a number of mirrors and picture frames secreted among the foliage, and a lilac tree is decorated with kitchen utensils, such as knives and forks, cheese graters and whisks. Recently, a bamboo beaded door curtain became redundant and, since it was already organic and indeed almost 'jungly', I fixed it up by the garden shed where it hung down and masked the dustbin, while also looking as if it shrouded a mysterious portal through the fence into a wild and magic land. Or something like that. Anyway, I liked it. But someone or something else didn't.

Temper, temper

Already mourning the desecration at the pond, I was now further appalled by an act of blatant vandalism. My beaded curtain had not only been pulled down to the ground, every single strand had been pulled off, and most of them were twisted and tangled, as if they had been shaken in a fit of temper, violent enough to also leave a scattering of bamboo and beads all round the shed. But though vanishing tadpoles puzzled me, destructive delinquency didn't. I knew instantly who was guilty. This was the work of Foxy.

There's gratitude for you

Heaven knows, I love to see urban Foxes in the garden, but it has to be said that our local representative is a bit of a hooligan. In recent times, he has chewed a gnome in half, beheaded a plaster pig, disembowelled a plastic Lapwing and pulled one of my water pumps out of the pond and gnawed at the flex. I trust he is grateful that I apprehended him before he electrocuted himself, but obviously not, or he wouldn't have destroyed my beaded curtain. Or eaten my tadpoles.

Dining at Oddie's

OK, I am not 100 per cent certain. I wouldn't risk taking him to court. But I have now heard from a few other people who have witnessed such a thing, and it does makes sense physically, as it were. Think of Foxy taking a wander through the Hampstead gardens, feeling peckish, imagining the well-healed householders picking at their Weight Watchers vol-au-vents and sipping their Pinots. 'I could do with a nibble,' he's thinking. 'Nothing too heavy. And a wee sip of pond water would go down nicely. I know, Oddie's.' So he slinks across a couple of lawns, leaps up on the Ivy-covered fence, trots along the top as daintily as a tightrope walker, and leaps off into Oddie's. Only to find himself trussed up in a bamboo curtain, which he can only disentangle himself from by tearing it to shreds. Not a lot of things spook old Foxy, but for a second there he thought he'd been trapped in some kind of net, and – own up – he panicked. 'Blimey, I need a stiff drink. Oddie's pond water! Not out of that horrible pretend pond; you can taste the plastic. The littlest pond. He fills that from the water butt. Pure rainwater, none more natural.'

From a sip to a slurp

So he takes a sip, and finds his nose covered in jelly! Maybe he's not sure what it is and tastes it and quite likes it. Or perhaps in Fox world frogspawn is a bit of a delicacy. My pond is like a huge bowl of caviar, or an enormous dish of tapioca or sago. There are little black wriggly things in there, too, not much meat on them, but quite an intriguing flavour. 'Reminds me of frog. Oh, of course, silly me – frogspawn. An acquired taste, but – you know what? – I think I have just acquired it.'

Whereupon, he slurps the lot.

Or was it?

Well, that's my theory, and I am sticking to it. The better news is that the tiny tadpole team that I decanted into the fish tank are doing fine, and some are looking distinctly froggish. And another glimmer of good news. Maybe Foxy is innocent after all, because I have just seen another possible culprit crouching on a garden chair and swiping his paws at the tadpole tank, which he won't get into 'cos I have put a lid on it. 'Come back for more, have you, Timmy?' Next door's cat! I know he had my Blue Tits, so I bet he thought: 'I'll have his tadpoles too!'

And I can't even do what I usually do to him. Flippin' hosepipe ban!

BLOG THREE

Whodunnit?

Suspense

OK, I'm sorry, I'm sorry, I'm sorry. Six weeks since my last blog. Six weeks waiting and wondering if my few remaining tadpoles have survived. How could I keep you in suspense like this? Don't I realise the pain of unrequited anticipation? Yes, yes; I do, I do. I felt the same about the next episode of *Homeland*, and *The Bridge*, and Manchester City's horribly inevitable last-minute goal. I even felt a twinge of tension before Pudsey, the dog, was quite rightly acclaimed as the most talented thing on television. And talking of television, I can't even use work as an excuse for not blogging. Well, not working on telly anyway, or doing anything I get paid for, well not yet anyway, though I am owed a small fistful of fees by the BBC and Channel 5 and half a dozen theatres. But I have been very, very busy.

Oh yeah! Doing what?

Well, there were those half-dozen gigs – oh ok, 'shows', talks, 'evenings with' type things. The 'with' being me and my longtime BBC TV producer Stephen Moss. Well, he isn't just mine; he's done lots of stuff on his own too. Anyway, we treat the audience to an evening of the 'best of' wildlife clips from over 10 years' worth of my own natural history series, introduced by and garnished with anecdotes and revelations. It is called 'Bill Oddie Unplucked', which means nothing but is further evidence that I wish I had been a musician (as in 'Unplugged', geddit?) It went rather well and we will probably do it again, so watch this space, as they say. So, that is a week 'on the road' accounted for, plus a couple of days recovering for me, while Stephen wrote another couple of books, no doubt now available from the appropriate website or even a real bookshop.

Tweet tweet

Writing is something else that has taken up quite a bit of my time. I have been tweeting with a regularity and occasionally an enthusiasm that has even surprised me. Anyone who has been following me will know that even my obsessions show little consistency, and the range of topics and targets could be described as 'scattergun'. Anything from the Queen's Speech to leaf blowers, via recent music releases, sport and, of course, conservation and animal rights.

My Twitter page is called Bill Oddie Official, to distinguish it from the disturbingly large number of tweeters pretending to be me. OK, it's maybe only two or three but it's still weird.

BBC

At rather greater length than 140 characters, I have produced two more monthly pages for *BBC Wildlife Magazine*, composed a letter from PETA to the Beeb protesting at the *Great British Menu* featuring recipes using foie gras (which involves hideously cruel force-feeding of geese, and in any case is not British). I also wrote a foreword for a forthcoming book called *The Dog That Survived the Titanic*, featuring heroic birds and mammals, both wild and domestic. It is hardly erudite but it is fun.

Daisy Daisy

During the time I have not been performing, writing or watching Chelsea do unlikely things to teams that are obviously better than them (but congratulations, especially to Big Drog) or keeping up with Nordic thrillers or English talent shows (not so thrilling), I have quite often ventured outside; proper outdoors, that is. The most distant excursions were to Norfolk. First, to the splendid Pensthorpe (for two years the home of *Springwatch*) for a delightful day celebrating the local River Wensum. I was accompanied by my new friend Daisy, who is yellow and has instant rapport with little children, even though they have not met her before. I am not sure exactly what species Daisy is. She says she is a Nature Bug, but as one precocious infant announced: 'She is not real!' Mind you, this didn't stop him wanting a cuddle. Daisy is a dispenser of awesome hugs which the kids accepted gleefully. Unlike my daughter Rosie, who at the age of four was introduced to Mickey Mouse at Disney World and instantly screamed and dived under the table at the 'Character Breakfast'. By the time she emerged it was time for the Character Lunch! I admit Rosie had my sympathy; my instinct was to do exactly the same. Out-of-work actors in character costumes with enormous heads but no voices give me the creeps. Daisy is not like that at all. Check her out. Check us both out.

Happy birthday

After the Pensthorpe day I drove up to Cley on the north Norfolk coast in the role of a sort of roving ambassador for the Wildlife Trusts, which were 100 years old in mid-May. That stretch of coastline is one of the most impressive demonstrations of what organised conservation can achieve. From the Wash to Cromer is in effect one continuous nature reserve, with various segments owned and managed by different but cooperating groups, such as the RSPB, the Norfolk Ornithologists' Association, Natural England, the National Trust and the Norfolk Wildlife Trust. Ah, if only the whole British coastline was protected like this! And the rivers, and the sea and – oh everything!

Dream on.

Simon, Sir David and me

The Trusts' birthday celebrations continued back in London. In the morning, at Gunnersbury Triangle, an urban reserve threatened by further encroachment of blocks of flats. That was the serious side of the day. In the evening, came the fully fledged 'posh do'. It was held at the Natural History Museum in South Kensington (maybe my favourite building in London). In the Great Hall, a massive towering dinosaur skeleton stands astride the orderly phalanxes of sparklingly coordinated round tables, bedecked in black and silver and each one sprouting a bouquet of wild flowers at the hub, which we were allowed – nay encouraged – to take home. The 'we' consisted of representatives past and present of sundry departments of the Wildlife Trusts and other NGOs (non-governmental organisations, i.e. charities). Neatly staged in natural breaks during the meal, we were treated to a sepia-tinted slide show of 'the old days'. Followed by words of wisdom from Sir David Attenborough (I don't think he knows any other kind of words!), a rallying cry by the current president, Simon King, and a quirky if perhaps slightly puzzling declamation from me of a letter I had been asked to write for my grandchildren to read in 50 years' time! By then they would be in their sixties. I just wouldn't be. A couple of people later told me that they found my performance 'inspiring'. Maybe because it proved it is possible to take a totally confusing concept and make it sound as if it makes sense. Perhaps they were politicians.

Talking of politicians

Guess where I was on the evening of Tuesday 22 May? The Chelsea Flower Show? No, I was there in the afternoon. In the evening I was at the Houses of Parliament! No, not sitting on the pavement with a placard. Nor haranguing the House during Prime Minister's question time. To be honest, I don't know what the bit we were in was called, but it took clearing several security checks and clattering down endless stone corridors to get there. I do know that if you go out on the balcony you overlook the Thames, which was nice. I had been asked to arrive at 8.30pm-ish, by which time I assumed that the MPs would have been let out of school and would now be guzzling and gorging in the tuck shop. Actually, I passed a whole corridor of 'tuck shops', or small dining rooms, each one harbouring a table groaning with wine, spirits and food, with a clientele of gentlemen groaning from having gobbled it all down too quickly. Meetings, receptions, committees, gatherings, piss-ups, jollies? Whatever they were, one component was blatantly almost missing. Women. Happily, this was not the case in our room. I am pleased – nay smug! – to be able to say that gender equality is exemplary in the world of conservation and animal welfare. If anything, I dare say women predominate, which I consider totally a good thing.

Cruel sports

Unfortunately, the other side tend to be largely men. What do I mean by 'the other side?' I can best elucidate by telling you more about that particular evening. The event was convened by The League Against Cruel Sports. It was to celebrate the appointment of a team of 10 investigators, whose task it is to track down people breaking animal-rights laws, and gather enough evidence to lead to convictions. This is what one might call the dark and dangerous side of the sort of work done by the RSPCA. I could recount distressing details, but suffice it to say that there is an escalating amount of cruelty occurring in Britain, much of it under the sickeningly inappropriate heading of 'sport'. Despite the ban on hunting with dogs, clandestine meets still take place, with a degree of attendant savagery. Even more shocking, there is a revival of the sort of practices that one might have thought became extinct during the Dark Ages. Badger-baiting, hare-coursing, cock-fighting and dog-fighting. The activities are, of course, illegal, as well as appallingly cruel. The perpetrators are also involved in betting on the outcomes of the fights, and also in many cases in drug-dealing, robbery and so on. These people are criminals. They are also almost without exception men. The League Against Cruel Sports operates on the front line. Other organisations fight their battles with political weapons: eloquence, argument, integrity, honesty, legality, unity of purpose and perhaps – most important of all – public opinion.

Cull, destroy, trap

Ironically, just as politicians are our allies –'our side' – so too are they often our enemies – 'the other side'. That evening, a small contingent of MPs supporting the Countryside Alliance had duly announced their presence with the sort of heckling and jeering which seems to constitute the language of debate at Westminster. It sounds more like a farmyard than an assembly of honourable gentlemen. Rather appropriate I thought. By the time I gave my speech the room was much quieter. Maybe they'd been taken home to bed, or maybe they didn't think I was worth heckling. Actually, I was rather disappointed not to have the opportunity of talking to some of the other side. I wanted to know why this Government is still minded to support the culling of Badgers. And why they are granting estate owners the right to destroy Buzzards' nests and to trap and remove the birds, illegal measures sanctioned in the name of protecting the pheasant-shooting industry.

And?

I could go on. I already have done. No doubt I will again. Meanwhile, if you are curious or incensed enough to want to find out more, you know what to do: visit those websites. Listen to the 'fors' and the 'againsts'. Examine your own beliefs and emotions. And ask yourself one of the most fundamental questions in life and folk song: 'Which side are you on?'

What's been eating my spawn?

Oh blimey! I have just realised! Here I am, utterly immersed in a swamp of contention and lost in a labyrinth of polemics, and I have forgotten to answer the one question that has been keeping you on tenterhooks for six weeks of sleepless nights. ***What happened to my tadpoles?*** For anyone who does not have the story indelibly and permanently emblazoned on their consciousness, may I suggest that you click on my last blog (yes, I know it was the middle of April, sorry, sorry) and remind yourself of the story. For those who can't be arsed, I shall rapidly reiterate for you.

Small pond in my garden. Overflowing with frogspawn. One morning, all gone! My theory: a Fox had swallowed the lot. Sort of Fox sushi. However, I had rescued about a dozen taddies and put them in a fish tank on the garden table. For a while, doing fine. Then one morning, fish tank devoid of life. Taddies on table nearby, dead or dying! What did that? Surely not a Fox! Surely a bird, but why didn't it eat them? Sheer devilment? Which bird? Chief suspects, all seen on table now and then: Magpie, Jay, Great Tit, domestic pigeons, Wren, Blackbird, Robin. Only one species actually caught in the act of tadpole dipping. The Robin! But surely he didn't have all the spawn as well? Expert comment by Simon King: 'I can believe the Robin, but not the Fox.' So, Simon, what *did* take all the frogspawn? 'I honestly don't know,' said Simon.

That is the mark of true authorities – when they don't know something, they admit it.

Politicians take note.

BLOG FOUR

Weather Report

Seen it, hate it

I feel as if I ought to write about the weather, 'cos everybody else does. Of course by the time you read this we could be sloshing around in floods caused by a sudden thaw, but for the moment the view from my window is as bleak, white and icy as the Arctic. It's not the least bit pretty, and it wasn't even after the first fall, which was about a week ago although it feels more like a year. In fact, it is as if January's weather has linked up with December's and the whole of spring, summer and autumn never happened. At least I enjoyed it in January. I did slip over three times on Hampstead Heath's notoriously treacherous pathways and cracked a rack of ribs – mine, not from the butchers – but I still had a backdrop of a white wonderland and an audience of some of the biggest and best dressed snowmen I have ever seen. This year, though, there are no snowmen, and everyone is trudging around with a doleful expression that says: 'Snow? Seen it. Hate it.'

Flustered

At chilly times like these it is customary to ask 'what about the wildlife'? Any day now the RSPB will be exhorting us to 'feed the birds, and don't forget to provide unfrozen water', while the classier Sunday papers will carry artfully photographed snowscapes, with shadows, silhouettes and deer or ponies. At least, hopefully, we will no longer have to stomach seasonal scenes of stampedes of scarlet-clad 'country folk' eager to match the red of their coats with the blood of a terrified Fox. Talking of which, I have just witnessed something rather striking in my garden. No, the Hampstead Hunt hasn't just galloped through, but I did see their quarry. At midday, in broad daylight, a truly handsome fellow peered down from my shed roof as if he had just that second risen from beneath a blanket of snow as white as his own chin. His nose shone, his eyes twinkled and his coat glowed so brightly it looked as if you could warm your hands on it. If only. Even as I made a flustered grab for my camera, he popped down out of sight, and popped up again on our next-door neighbour's shed roof, which – unlike my flat one – has a steep slope, down which Foxy duly slid. He dropped – well, OK, tumbled – behind a privet hedge, hopefully in a nice soft snowdrift.

Dangle

I have yet to notice any of my garden birds looking unusually hungry. This is because they are continuously gorging themselves on the three kinds of high-class birdseed I put out for them, or on the top grade peanuts, or the extra-nutritious fat balls, or the three bowls of live mealworms. Not that they look terribly alive in these temperatures. I wonder if mealworms feel the cold? Never mind, they'll soon be snug and warm inside a Blackbird's tummy. Or a Song Thrush's, or a Robin's, or any one of three species of tits, or a Jay's or a Magpie's. Jays and Magpies aren't supposed to dangle. Especially not on small plastic feeders with conical lids on them to deter larger birds such as, for instance, Jays or Magpies, both of which have in my garden learnt to dangle almost as deftly as the Blue Tits to get at the mealworms. Mind you, it took them a bit of practice before they got the idea, and a lot of frantic wing flapping and falling off, which I have to admit gave me hours of enjoyment. Let's face it, Jays and Magpies are such cocky blighters they could do with losing their dignity now and then.

Clammy

So how are the rest of my garden critters reacting to the mini ice age? Foxy looked fine. The squirrels are, if anything, even more hyperactive, especially when they leap onto a snow-covered branch and give themselves a cold shower. The effect is even more spectacular when they jump straight onto a hanging bird feeder so that their weight pulls on the wire, releasing a veritable avalanche of snow from the tree above. It's like one of those rope and bucket showers intrepid campers set up, only they are probably in a clammy jungle, not a frozen garden.

Scare

The local Wood Mice really don't care for this weather at all. I know there's at least one living in the shed, because I have been putting out a bowl of seed, which is duly eaten every night. I know it's a mouse because of the teeny black droppings around the bowl. Also because I had a right old scare the other day when a plastic bag started to rustle and move across the floor, and when I plucked up courage to touch it a Wood Mouse literally leapt out vertically as if it had been fired from a catapult. The shed is not warm. I reckon the spring-heeled leaper really is the only mouse in there. Why? Because the rest of them are inside our house!

Goodies

It has happened before. Even at times when it hasn't been so cold. In fact, I am not convinced that it is the inclement weather that causes my Wood Mice to move into our living space. I think they just like it. I know they enjoy shredding the paper covers off my irreplaceable collection of Goodies' vinyl singles, which I had understandably banished into the corner of a dark cupboard. Just the sort of place mice love. Another one, or possibly two, completely stripped a pink fluffy plant-pot cover my wife had been given as a birthday present. She had not been grateful to the donor, but she was grateful to the mice for ruining it. She also loved the idea of a brood of baby mice being raised in a pink fluffy nest.

Unsavoury

However, the mice's favourite bit of our house is what we call the TV room, because the only thing we do in there is watch TV. It is quite a big TV (HD but not 3D), in quite a small room, but there is just enough space to accommodate a large cosy couch and a slightly dilapidated armchair. The armchair is mine. A detective could tell it belongs to a man of grandad age, because the carpet in front of it is stained with blotches of spilt wine and takeaway curries, plus a scattering of crisp crumbs, and the occasional grape or blueberry. This unsavoury – or maybe savoury – detritus is no doubt the mouse equivalent of a help-yourself buffet, so it's not surprising they are frequent visitors. Not that I ever see them for long.

Climax

Their appearances are almost subliminal, but their timing is ruthlessly impeccable. There I will be, slouched on my greasy yet comfy

I KNOW I DON'T BELONG IN HERE..
BUT I CAN'T GET BROADBAND IN THE WOODS.

cushions, enjoying *Match of the Day* or *Live at the Apollo*. A climax is nigh. The centre forward is bearing down on goal, or the 'stand-up' comic is just about to deliver his punchline, when… zip! What was that? I am distracted for barely a second, but by the time I turn back to the screen, the ball is nestling in the back of the net or the audience is screaming with laughter. As is also, I suspect, the mouse. I know he does it on purpose. He hides under the settee, checks out what's on the telly, waits for a vital moment, then propels himself across the hearthrug and vanishes behind the waste-paper basket. There is no point in me creeping over and peering round, under or in the basket. He won't be there. And he won't be anywhere else I look either. Frankly, I could strip the whole room of every piece of furniture, and I still won't find him. Like as not, I won't even find any potential mouse holes that he may have escaped down either. A magic mouse? I needed to know. So…

Paranoid

Late one night last week, I missed a vital wicket in the Test Match because just as the bowler was running up, the bottom of the curtain moved and I looked away from the screen. The triumphant cry of 'Owzat!' seemed aimed personally at me. I imagined a gleeful mouse, hiding behind the curtain, revelling in his mischief. 'Ha-ha. Gotcha Oddie!' But hang on. Was it a mouse that had got me, or my own imagination? The fact was, I hadn't seen anything at all. Except the curtain moved. Or did it? Was I becoming paranoid? Was it a hallucinatory mouse, or a real mouse? Before I retired for the night, I set a small humane trap – baited with Haith's finest birdseed – right in front of the telly.

We shall see

So what did I find the next morning? A cute little mouse! I admired and photographed him (or her), and released him in the street's communal garden. About 100 metres down the road. 'Not far enough,' several regulars at our local coffee shop have told me. 'You have to take them at least a mile away, otherwise they'll just come back.' Well, we shall see, because before he leapt to freedom, I anointed this little fellow's feet with green food dye. He is a marked mouse.

Watch this space.

BLOG FIVE

Mine all Mine

Hey ho, hey ho

I'm sorry, I'm sorry. I know I haven't blogged for several weeks, but I was in no fit state. Actually, I was in Zambia, East Africa. Oh yeah, swanning around the luxury safari camps in my lightweight linens, I dare say you assume. Well, as a matter of fact, for the first four days I was done up in overalls, wellies and a hard hat. Not the ideal garb when the temperature is high and the humidity is higher, but obligatory if you are dodging diggers, bulldozers and falling rocks, or seeking refuge down a mine shaft. A mine shaft where a large section of the tunnel ceiling was being contained from collapsing by what looked like a giant rope hairnet. In fact it *had* collapsed only a week before, but we were assured that the net meant that that bit was safe now. Of course, no one could vouch for the rest of the tunnel. There was much nervous banter about Chilean miners, being rescued by Lassie, and: 'Please can we get back above ground as quickly as possible?' This we did, only to find ourselves careering through a dust cloud, crammed into a rattling Land Rover, precariously descending into what looked like a massive volcanic crater but was in fact the work of the diggers and bulldozers that had gouged out this giant quarry. At the bottom, a small team of workers were scrabbling in the mud and staring intently downwards, as if all of them had coincidentally lost their contact lenses. The truth was that keen eyesight was essential to their task. They were searching for emeralds.

Emerald city

Now, I have to admit that gemstones and jewellery are not high on my list of interests, nor indeed do I consider them 'precious stones' because they look dazzling, but simply because they are worth a lot of money. Rather like a lot of contemporary art, the value of an emerald seems to be dictated only by what some extremely rich person is prepared to pay. Whatever the price paid, that is what the stone is worth. Until it gets sold to another rich dealer, whereupon it becomes worth even more. During the course of our visit we were frequently enlightened by lectures, demonstrations, PowerPoint presentations, and one-sided conversations about the complexities of the international gem market. To be honest, I understood barely a word. What I did grasp, though, was that the management of this mining company was doing well enough to be able to contribute a considerable sum of money to 'good causes'. In fact, they already were providing medical and schooling facilities for the local people, and – and here at last comes the reason for me donning a hard hat and braving an African mine shaft – they had also subsidised an elephant conservation project in India. They were now anxious to support their own Zambian wildlife, and we were anxious to let them. 'We' were a small deputation representing the World Land Trust, led by their Chief Exec, John Burton. To find out more about the excellent work they do please visit their website: www.worldlandtrust.org.

Show me the money

I have known 'purist' conservationists baulk at the idea of accepting funding from big industrial companies, especially ones that have the potential to destroy or ruin habitats. I remember 50ish years ago when large oil companies, such as Shell and BP, first began espousing wildlife and ecology by funding the publication of new bird books, and maintaining nature trails and reserves and so on, there was a widespread suspicion that it was 'guilt money', and the offers were not always accepted. To put it bluntly, the majority of conservationists nowadays take a more pragmatic attitude to industry, development, etc. The policy is: make sure they do as little damage as possible and hand over as much money as you can get! To be fair, many of these companies are genuinely anxious to appear both responsible and generous.

Trust the Trust

Of course, mining – for emeralds or anything else – is potentially a great wrecker of habitat, especially if it is open-cast. I well recall the state parts of Northumberland and Wales were left in after the ravages of open-cast coal-mining. Fortunately, nowadays there is an obligation to repair the damage. Indeed, excavation can be ultimately constructive. One generation's quarries can become the next generation's lakes, marshes and reedbeds. Just add water! This is already happening in Zambia at the emerald mine we visited. Further negotiations with the World Land Trust are continuing, even as I blog. I will keep you posted.

All Greek to me

All that sounds pretty constructive, doesn't it? So why did I start this blog off with a whinge? Ah, well, I am afraid our emerald-mining hosts made one mistake. On the last night they took us to a Greek restaurant. Not what you'd expect in Zambia, and indeed the food was not so much Greek as sort of Italian. Oh, let's face it, it was a pizza house. Instead of doing the prudent thing, like leaving, at least three of us chose dishes that involved fish or prawns. At some point during coffee, it was pointed out that we could hardly be further away from the sea, in a country probably not renowned for its rapid refrigerated transportation. The consequences were inevitable, unpleasant and lasted throughout the three days we spent at a delightful lodge in Luangwa. Alas, my condition was anything but delightful. A dawn safari drive is not the best thing to calm a queasy tummy and loose bowels. Erratic movement on rutted roads has much the same effect as being tossed around in a rowing boat in a squall, while being caught short leaves you with some pretty insidious choices. You can ask the driver to stop so you can nip behind a bush, but you risk antagonising a family of resting Lions. One can't imagine they would be pleased! Or you can try and hold it in till you get back to the lodge. You can try, but you may not succeed. In case you are wondering… you really don't want to know. And you certainly don't want me to describe it!

Would I recommend Zambia for an African adventure? Yes, yes and yes. Just don't eat the Greek pizza.

Tubby little singers

News from my garden. I am happy to say the Wrens are back. If indeed they ever went away. Certainly, a month ago there wasn't a Wren to be seen or heard in my neighbourhood and I feared that they had been hit very hard by the coldest part of the winter. Then, at the end of March, it was if someone had switched on 'Radio Wren'. Now there must be half a dozen tubby little singers belting it out on what was previously the silent circuit. So did they go, or just go quiet? Some of our birds do perform what are known as 'weather movements' if the weather gets too nasty. Lapwings, Skylarks, some pipits and finches fly off south-west and maybe even over to southern Ireland where it is milder. But Wrens? Surely not? Maybe they have learnt how to hibernate. Anyway, I am very happy they are back.

Shock!

As it happens, I was recently browsing the results of the RSPB's Big Garden Birdwatch. I noticed that Wrens were down at number 20. A considerable slide for a bird that has been regarded as Britain's most numerous. The top two also surprised me: House Sparrow and Starling, though that doesn't mean that they are increasing in number. I have seen neither in my garden for years. Another shock result, no Blackbird in the top 20! And Blue Tit appeared twice at both three and four until the RSPB scorer corrected the mistake and inserted Blackbird at number four where it belonged. Phew! Just in time to head off a 'Save our Blackbirds' campaign on Twitter.

Right then. I am due to cross the Atlantic in a fortnight's time. Going to South Carolina to meet some gibbons. Honestly. All will eventually be revealed. The only downside is that I will miss The Wedding. It will serve them right for not inviting me. They seem a nice young couple. I wonder if she feeds the birds instead of shooting them? It would be great to see a royal setting a good example wouldn't it? After all, it is the *Royal* Society for the Protection of Birds.

BLOG SIX

Blow me Down

It's a twister!

Blimey, doesn't time fly, especially when I have been – flying that is. If I remember rightly – it's so long ago – I concluded my last blog by bidding you farewell before boarding a plane bound for Charleston, South Carolina, USA. There I would proceed to Summerville to visit the headquarters of the International Primate Protection League (IPPL), and also the residence of 30-odd gibbons. I was looking forward to it very much. The only small frisson of trepidation was triggered by news of extensive tornado damage in the south-eastern states of the US, including South Carolina. I scared myself even more by logging on to the American Weather Channel, which featured a continuous stream of video clips of ever blacker, ever bigger, and ever more destructive 'twisters'. These had all been sent in by viewers who invariably added a live commentary along the lines of: (please imagine a southern accent) 'I sure seen some twisters in my time, but damned if I seen one big as this!' Or – as if to top that – 'God damnit if there were not just one twister, but two of 'em! Like a pair of angry twins coming straight at us. I'm thinking, Lord, between 'em they gonna lift up the house. All we could do was sit tight and pray. So we did. And you know what? Just before they hit the house, them twister twins split either side of us. One of 'em clean lifted my wife's Chevy off the front drive, and the other one grabbed my pick-up from the back, threw it in the air, spinning like a top, and let it drop bang slap in the Reverend Millhouse's prize rose garden. And God damnit if there weren't four or five other vehicles already been pitched there by the twister, and they were on their sides, on their backs, wheels still spinning, like snapper turtles that had been tipped over by the kids. And I'm thinking, Lord, this is truly hell, when all of a sudden, the wind stops roaring and the rain stops pouring, and I figure we are right in the eye of the tornado, 'cos the sky is all still and calm and empty. No more pieces of fenders and trash cans and picket fences flying through the air. Nothing. Except, God damnit, I do believe it started to snow! But, of course, it weren't snow, it was rose petals! Floatin' down, pink and red and yellow and... Well, it was like it was showering down bits of rainbow. I gotta tell you it was beautiful. And now I'm thinking, if this ain't hell, maybe it's heaven. Only one thing spoilt it. Suddenly Reverend Millhouse comes out of his house, sees what's happened to his prize roses, and boy is he pissed! He starts cussin' and hollerin' and yellin' right up at the sky, like he's complaining to the Lord himself. I swear he called him a "holy mother f*****!" Sure weren't words you expect to hear from a man of God.'

Very scary

I admit I have embroidered that account a wee bit, but I promise you, there was some very scary footage of twisters and the appalling damage they cause. Nevertheless, I have to admit that a little bit of me was rather hoping I would get to see a twister myself. From a safe distance.

But I didn't. During the eight days I was in South Carolina, I experienced two light showers and some morning mist. Otherwise, it was blue skies every day.

But I did see some gibbons

The IPPL sanctuary is not a zoo. It is not open to the public. It is simply a safe home for some gorgeous animals that have previously lived in inappropriate or indeed horrendous conditions. They are 'rescue' gibbons. Some spent years being used for 'research' in laboratories. Others were saved from incompetent private owners or substandard zoos. They can never be returned to the wild, but at least at Summerville they are safe, cared for, and, most clearly of all, very much loved. To get to know them better and to learn more about primates and their problems I recommend visiting the IPPL website – www.ippl.org – and also do a bit of gibbon Googling. There are some great photos, but believe me, you haven't really experienced the best of a gibbon until you've seen it and heard it swinging and singing.

Buzz

Meanwhile, back in London. Something completely different, and to me completely new, though very close to home. Very close. In a tree just outside the back door. It is a slightly straggly lilac, which has become an extension of what I call 'the magic tree' in that I have chosen to festoon it with a kaleidoscope of wind chimes, fragments of coloured glass, beads, fake butterflies and an ornamental bird-box that is painted yellow with a red roof, and has never been used. But it has now. But not by birds. A small colony of bees has moved in. This sort of thing has happened before in my garden. I was once cleaning out a box and got stung by a 'solitary bee' that wanted to be alone. Another year, the same box was commandeered by a few White-tailed Bumblebees, which are quite tubby and slow and bumbly and therefore not too hard to identify. They are not, however, the only species that has a white tail. For instance, the bees in my magic box. They appeared to also have a largely black body and a gingery shawl (yes, I know, I should use all the scientific words for insect parts but I can never remember them). In fact, I was having trouble getting a decent view because they wouldn't keep still. Every now and then, one would zoom in or out of the hole in the nestbox, but the most conspicuous action was a small party that just kept swirling around outside the box, bouncing, dipping and diving, as if they were dancing. A rather unruly square dance perhaps?

Let it bee

Next step, consult the literature. In my pre-computer days I was pretty addicted to the excellent laminated identification charts produced by such organisations as the Wildlife Trusts. I also have a few favourite books on garden wildlife. So I searched. But I could not find my mystery bees.

However, it just so happened that later in the week I was due to attend an event at which Buglife would launch its 'Get Britain Buzzing' campaign (for more on that visit www.buglife.org.uk). I had barely entered the room when my eyes fell on an illustrated leaflet about 'bees in your garden', and there, among other species I knew well, was a photo of my bee! It was labelled Tree Bumblebee. Never heard of it! But my bees were in a tree. I soon collared a bee expert – not difficult at Buglife! – and quizzed him about Tree Bumblebees. What he told me was instantly conclusive. 'They often nest in old bird-boxes.' Like mine. 'They have this behaviour of dancing outside the nest.' Exactly like mine! 'They arrived from the Continent about five or six years ago.' Which explains why they are not in the books. 'Any problems with them?' I asked. 'No. In fact it's good to report that we have another pollinator in Britain. That's what this Buglife campaign is all about.' Exactly. So, I've got Tree Bumblebees! Welcome.

Dead parrots?

Another thing I've got in my garden is parakeets. There are, of course, parakeets all over London and beyond and there have been for years. There are probably more than 50,000 of them out there in the wild and they are now accepted as British birds.

It has therefore caused a right old kerfuffle that the government, with the support of the RSPB, has decided to 'cull parakeets'! Thus ran the headline in several national newspapers a few weeks ago. The article was often accompanied by a picture of a parakeet, just like the ones in my garden. The bird in question is a Ring-necked (or Rose-ringed) Parakeet.

On the face of it, this is a pretty unbelievable decision. Kill more than 50,000 birds! That's not a cull, that's slaughter. And it is surely impossible. Added to which, people – well most people – have grown rather fond of the old Ring-neckeds. And the RSPB approves! Outrageous!

Get it right!

Hang on, hang on! Remember the irrefutable adage by which we should all temper our opinions: 'Don't believe everything you read in the papers.' In the case of some papers, this could be amended to: 'Don't believe *anything* you read in the papers.' To which I would add: don't assume that they've got the right pictures in the papers. I sadly report that at least two so-called 'quality' papers proudly printed photos of Ring-necked Parakeets, alongside the 'parakeets to be culled' story. Wrong!

Monks' menace?

The proposed cull refers to Monk Parakeets, which are quite different from the Ring-neckeds, have a much smaller population and potentially do far more damage. There are thought to be about 100 to 150 birds on the loose (mainly round London) and the problem is that – unlike the Ring-neckeds, which live in tree holes – the Monks build an enormous nest of twigs and branches, which may be in a small colony, so that half a dozen nests meld together in a huge clump which can be the size of a small car! Unfortunately, this construction is often perched on a telegraph pole or electricity pylon and woven into the wires. In the US this has been known to cause blackouts.

I am not going to comment any further, but in case you were about to convene a Save Our Parrots campaign, please make sure you've got the right one. And don't depend on the papers for the facts. Although sometimes…

Chinese crackers

Did you see the story about the Chinese watermelon farmers who overdosed their crop on a chemical 'growth accelerator' and arose one morning to find their fields erupting with hundreds of exploding melons? What a wonderful image! It's at times like this I wish I was still doing a comedy show!

BLOG SEVEN

Shall We Dance?

Strictly

I could have been Ann Widdecombe. Or am I more Russell Grant? So what weird fantasy is this? Have I decided to come out and confess my secret desire to be on *Strictly Come Dancing*? Well, the fact is I could've been. No, honestly, a couple of months ago the *Strictly* office rang me and asked if I would like to come in and discuss my being on this year's show. I decided to go into the office if only because it gave me an excuse to get all nostalgic about the old studios at BBC TV Centre, which are due to be sold off rather than plastered in blue plaques and preserved for ever as a shrine to the great BBC shows of yesteryear. There was also the small possibility that I might bump into one of the bosses who might recognise me and remember that I used to appear a lot on BBC Two, and could do so again. If they asked me. I can dream.

Pan's People

But here was a dream that could come true. Or would it be a nightmare? It was up to me. The people in the *Strictly* office were really nice, and it wasn't long before we were giggling together, as we acknowledged that I would be cast as the little fat one who couldn't really dance. Vanity forced me to inform them that I had danced on Broadway and won rave reviews, and had frequently been choreographed by the lovely – and sadly now late – Flick Colby, the mentor of Pan's People no less. On the other hand – or should it be other foot? – my style was strictly *not* ballroom. This was partially because the mention of quicksteps and foxtrots brought back painful memories of the Sixth Form Dance, where I was an eternal wallflower, sitting forlornly in my ill-fitting dinner suit, wishing that my bow tie would finally asphyxiate me and put me out of my misery.

Writhing

That's another thing. Costumes. Imagine me in a sky blue silk trouser suit, with a neckline plunging to the waist, revealing a belly that could compete with a Christmas pudding. Actually, I *wouldn't* imagine that if you want to sleep tonight. As for the combination of fake tan and facial hair! My head would look like a sweet chestnut. The chief interviewer lady attempted to shatter my objections by pointing out that back in the 1970s I had shamelessly cavorted in The Goodies in everything from a miniskirt to a mouse costume, and that silk trousers, hippy waistcoats and bling was my standard dress when I went clubbing. I couldn't deny that, but I pointed out that the dancing in those days consisted of little more than rhythmic writhing, which took place largely in the dark. 'The truth is,' I explained, 'I have never ever done any ballroom dancing.'

'Well, that is exactly the point!' the lady countered. I couldn't deny that either.

Perfect

To be perfectly honest, I could see all too well why I was an ideal choice for *Strictly*. I was totally unsuited. Perfect. What's more, we'd shared a nice pot of tea, biscuits and some merry banter and

I didn't want to disappoint them. 'You'll love it,' they assured me. 'Everybody who's done it has.' Maybe, but what about those watching? Like – for example – my family. I decided to give *them* the vote.

Hate

My middle daughter, Bonnie, who is a dance teacher and choreographer and therefore knows about these things, was practical in the extreme. 'It'll kill you,' she said. 'It's very hard work, and you are not very fit, and let's face it, you are a bit overweight.' My wife, Laura, who abhors negativity, immediately spotted a positive slant. 'You won't go on a diet, so if you lost weight that'd be good.' Then she added, almost as an afterthought: 'But not if it kills you.' She clearly felt it incumbent on her to come up with some of my more cerebral flaws. 'You are very bad at learning things.'

'You hate being taught,' agreed Rosie, daughter number three. 'When you tried to learn guitar for that BBC Two programme you made your teacher cry!' Kate – daughter number one – agreed totally. 'You'll say what you think and upset someone, and then you'll say you hate ballroom dancing anyway, and all the viewers will be thinking "so why is he doing it then?", and they'll all hate you too.'

OK, I'd got the message. 'So I'll take that as a "no" shall I? Kate, no. Rosie, no. Bonnie, no. And Laura?'

'You must do what you want to do. But … no.'

'So, that's four nos!'

Even as I spoke those words, the truth dawned on me.

'You know what? I want to watch *The X Factor*.'

Celeb

Oh dear, half a blog gone and I haven't mentioned wildlife. I promise I will. Meanwhile, let me tell you why I turned down *I'm a Celebrity… Get Me Out of Here*, twice. As it happens I reckon this is one of the 'best' of the 'reality shows', in so far as there has been some genuine interaction between an unarguably diverse bunch of people, and nobody could deny that it is physically and psychologically challenging. The costumes are of course much more my style, and even without my binoculars I would surely spy some interesting Australian rainforest fauna. Indeed, when I was first approached, somebody commented: 'It should be easy for you, you go to that sort of place all the time. Just carry on watching the wildlife.'

Cockroaches

Watching it, yes, but not eating it. Call me a soppy old insect lover if you like, but chewing live grubs and grasshoppers, or crunching through barrels of cockroaches, is not my idea of animal welfare. Maybe all the contestants should be forced to watch Attenborough's *Life in the Undergrowth*, before being handed a plateful of beetles to swallow. Insects are not some kind of lesser life. Here's a thought: the most malevolent and the most beneficial wild creatures in the world are insects. Take malaria-carrying mosquitoes and pollen-carrying bees. Bad or good, I find the idea of any living thing being sacrificed for laughs on a TV show – what's the word? – distasteful.

Serious

Talking of dodgy food, it is this time of the year that weekend newspapers and colour supplements are likely to carry articles on gathering, cooking and eating wild fungi. One word of advice: don't! The article will no doubt include a warning along the lines of: 'Make absolutely certain that the mushroom you have picked really is edible.' Yes, and how do we do that? Look it up in a mushroom identification book, or on the Internet, I presume. The warning further informs us: 'Many poisonous fungi look very similar to an edible species and vice versa.' If you want to confirm that: also look in the book. Peruse the pictures – whether photos or illustrations – and you will find pages and pages of apparently identical fungi. Many are the same colour, though in some species the colour may vary. So too may the shape and size. Some can only be safely identified by dissecting them! Others, you'd probably have to eat to be sure what they are. It's either that one – 'edible when fresh' – or that one – 'causes severe stomach pains and vomiting, and can be occasionally fatal'. If you think I am just scaremongering or have recently joined the Mushroom Liberation Front, get a book or two, find some fungi and try and sort them out for yourself. But before you fry 'em in butter and garlic, remember a mistake could have serious consequences. I repeat my advice: don't!

No advertising

And finally… I am often asked which binoculars I use and which 'outdoor clothing' I wear. Such secrets must never be revealed on BBC TV – no advertising knowingly allowed – but I can tell you here.

I use Swarovski optical equipment, and I am kept snug and dry by Country Innovation. And neither of them is paying me to say that.

I am now to be found monthly in the sumptuous and vastly informative *BBC Wildlife* magazine. And they *are* paying me, but that's for writing the articles.

BLOG EIGHT

This Just in

I thought you were ill

Who was it said: 'News of my death has been greatly exaggerated?' Was it Mark Twain? Whoever it was, it was some time ago, so presumably it isn't exaggerated any more. However, I am beginning to know what he or she meant… As far as I am aware, I am yet to appear prematurely in an obituary, but I am gathering evidence that people are assuming I am in worse shape than I actually am. For example, my agent has reported that on more than one occasion when he has been peddling his client (i.e. me) to a TV company or a producer, he has met with the response: 'Bill Oddie? I thought he was ill.' The operative word in that statement is 'was'. Past tense. Back in 2009, I was indeed floored by a deep and dangerous depression, however – happily – illness was duly followed by recovery, which has – even more happily – not been followed by relapse, and – more happily still – this state of 'not being ill' has been maintained for more than two years, as friends, family, neighbours, acquaintances and various medical authorities will testify.

Chunky blonde or quirky sweater?

During this time, I have done a fair amount of travelling, and have spent many enjoyable hours, days and even weeks in the company of those who work for the many terrific NGOs concerned with conservation, animal welfare, alternative energy, child care and mental health. I am proud to be able to help however I can. These people are truly inspirational, and a lot of fun, even though the dilemmas they deal with often aren't. I have also done quite a bit of writing, and have given interviews to several magazines and newspapers. I've done a lot of work in my little garden, listened to a lot of new music, watched a great deal of sport on the telly, become addicted to Nordic thrillers and *Homeland*, and developed old-age crushes on three heroines, though the quirky blonde in *The Bridge* does frighten me a bit. *The Killing*'s Sarah Lund I admire for her undistractable concentration, and her chunky sweater, which is very like the ones worn by fishermen in Shetland, which is one of my favourite places. I used to have one myself. *Homeland*'s Carrie Mathison (Claire Danes) is of course bipolar – as am I – so naturally I have empathy for her. And vice versa? If only.

Healing the well

Anyway, I put it to you that these are the activities of a normal, healthy man. I admit I do take a certain amount of medication, and I have occasional check-ups with my GP, but I no longer attend a shrink, nor belong to any kind of therapy group. I am not knocking them, I am simply pointing out that my health regime is much the same as that of anyone else: get some exercise, lose some weight, cut out the chocolate and cut down on the booze. One might call them remedies for the well people. So, that is my message to anyone out there who might be curious, concerned or indeed totally apathetic: *I am not ill.*

I hope you are all OK too.

Self-interest

Meanwhile, what's going on in my garden? A veritable Robin fest, that's what. In spring I had a pair of Robins trusting enough to take mealworms from my hand. This was not only entertainment for my granddaughters – and for me – it was also my way of hopefully encouraging them to nest in my Ivy, or in one of my nestboxes. Call it bribery and self-interest if you like, but it was also altruistic, as I tried to assure them each morning. Yes, I talk to the Robins. 'You guys nest in my garden and you'll have enough mealworms never to have to bother rummaging around looking for caterpillars or tiddly little worms. You'll be able to feed your chicks, and yourselves, without flying more than a few yards from your nest. Surely that's an offer you can't refuse?'

Not natural

But it was. As the soggy weeks of spring passed by, I had to choke back the pang of disappointment at seeing the pair gathering moss from my rockery and flitting over the fence with it. I was forced to face the rather hurtful fact that my Robins were nesting next door. This was particularly ungracious, since I would not call my neighbour a natural nature lover. (He nags me nearly every day: 'When are they going to shoot those bloody parakeets?')

Knackering

However, after a couple of largely Robinless weeks, suddenly things got a bit livelier. I knew instantly what had happened. The egg had hatched. And where did the parents immediately go in search of food? My garden! I do not hold grudges and I was not bitter, but as I tipped extra mealworms into my feeders, I could not resist letting the faithless parents know some of my feelings. 'Oh hello, so who's hungry then? You two? I expect it's because you're so busy feeding those babies, you don't have time to feed yourselves. How many have you got? Three? Five? The last Robins that nested here – in *my* Ivy, as it happens – had four. Course it was no problem. All mum and dad had to do was flit down onto the feeder – well three feeders actually, all with mealworms – so they could keep the kids full up, take a break while they digested their worms, and have a snack whenever they felt like it.

Of course, I would've done the same for you, but apparently you prefer to live next door, where there are no bird feeders, but there is a cat, and that bloke ranting on about the parakeets, and you have to keep flapping backwards and forwards over the fence. Yes, I am sure it *is* knackering, but it's what you wanted. Anyway, perhaps at least you'll bring the kids to see me when they fledge.'

They're leaving home

But Robins – and most small birds – don't work that way. Yes, you've seen *Springwatch*, when there has been a family of Blue or Great Tits lined up on a branch, and they shiver their wings and one of the parents feeds them, but it doesn't last long. It is essential that young birds learn to feed themselves as rapidly as possible. If they are reluctant, mum or dad is likely to give them a push, quite literally. No good protesting: 'I can't fly very well yet.'

'Yes you can, I've seen you. Now off you go, before I give you a good pecking.' Almost overnight, maternal affection turns to belligerence. Perhaps we humans should try it when the teenagers won't leave home.

Irresistible

Now, let's fast forward to just over a week ago. I go out in the garden for my morning routine. Fill up one feeder with sunflower hearts, and another with peanuts (both equally irresistible to finches, tits, squirrels, a Great Spotted Woodpecker and a dozen parakeets). Then, I put handfuls of mealworms in four plastic feeders: two hanging and two stuck on the window. These are meant for little birds, but neither Magpies nor Jays are beyond dangling and contorting themselves and gobbling up greedy beakfuls. I often lie on the couch by the backroom window, listening to records with my eyes closed. Believe me, it is quite disconcerting to be woken by tapping and scuffling and to look up to a Jay apparently trying to peck through the glass inches above my face.

Little visitor

The final task of the morning round is to check my moth trap. This is a very basic 'light and box' affair. The night had been damp and chilly. I didn't expect much and there wasn't. But no sooner had I chivvied a Willow Beauty onto a little plastic jar and put it on the garden table, than I became aware of a little visitor perched on the nearby railing. A newly fledged Robin. A riot of spots. A fluffball with measles! It couldn't yet fly too strongly, but boy could it hop. It hopped straight towards me. It hopped onto the table and pecked at the moth in the jar, but – just like at the supermarket – the food was sealed in plastic so rigid it was unopenable. Fluffball cocked his or her head towards me, then hopped off the table and onto my foot!

Good behaviour

Since then he has perched on my lap and on my shoulder. When I hold out a handful of mealworms, he hovers above it for a moment as if selecting the one he fancies, then grabs it in his bill and retires to a shady spot, swallows it and comes back for more. At the risk of confessing to soppiness I admit this little bird cheers me whenever we meet; his behaviour – as the scientists say – also intrigues me. Why so tame so quickly? Last year, I managed to get an adult Robin to take food from my hand. Could that have been one of Fluffball's parents? Could tameness be hereditary? He certainly hadn't had time to learn this behaviour by copying either a parent or me.

There's more

A few days later, not one but two adult Robins appeared in the garden. Mum and dad coming to conscientiously – or guiltily – check on their progeny? Or were they just hungry? I presumed the latter, as one of them swooped in for a mealworm, only to be immediately repelled by a flurry of wings, beak and feet and be literally chased away by its own offspring. There was no fight, no defence, no retaliation, and it happened with both adults. Conclusion: juvenile Robins show zero tolerance to their own parents. Interesting.

Or to their siblings. A couple of days ago I was wondering if I was seeing double. Or was Fluffball playing hide-and-seek? Or… Oh, the fact is that there is not just one young Robin in my garden – there are two. Plus mum and dad hop in now and then, and – for a couple of days – we had a visit from another adult in such a tatty state of moult it was probably embarrassed to be photographed. So that's a quota of five Robins. There has been no great violence, but there have been a few skirmishes, and the siblings are yet to show the slightest affection for one another. The nearest I have observed to a truce is when they perched on the same branch only about a metre apart, though it was raining heavily, so presumably they had agreed to share the shelter till it stopped.

Worry

Anyway, it was bucketing down again last night, and so far today I haven't seen any of them, so I'd best go and check that they are OK. That's the trouble, they may not be my kids, but I do worry about them.

WHO AM I?

EPILOGUE

Do They Mean Me?

Angry

Do you recognise this person? More to the point, do I recognise this person?

In the mid-1960s, he was quite an angry man and had to leave Rochdale due to violence and drinking. He ended up in a hippie commune in London. His metal (sic) health doctor encouraged him to channel his anger into humour. He still has a bad temper and was reported to have struck John Craven with a tripod over a disagreement about tea!

Who is it?

That's me that is, apparently, according to some website I accidentally visited the other day when I was idly passing a spare five or six hours Googling myself (all celebs do it). Most of it is simply inaccurate or untrue. The mid-1960s were one of the happiest times of my life, when I was appearing on Broadway with John Cleese and Tim Brooke-Taylor, and beginning a relationship with the lady who became my first wife. I had long left Rochdale. I left when I was six! Presumably not due to violence or drinking. Well, not mine anyway. The hippie commune bit has a smidgeon of truth in that during the 1970s a number of friends lived in a big house together. We had a cooking rota, but we weren't hippies, and I didn't 'end up' there; I owned it! I like the idea of a 'metal health doctor'! Heavy metal I presume, or was he a robot? Either way, I was never advised to channel anger into humour, though it is not bad advice, and may indeed have been what I subconsciously did do, but no one put it that way until I visited a psychotherapist relatively recently. As for still having a bad temper, that's not up to me to judge, but I have never ever even thought about striking John Craven, let alone with a tripod, and certainly not over a disagreement about tea!

Fight!

I would, of course, love to know where – or who – all this nonsense comes from, but what really intrigues me is what exactly was the disagreement? Were John and I squabbling?

'You're drinking my tea.'

'No I'm not, it's mine.'

'No it isn't.'

'Yes it is.'

Or were we arguing about what kind of tea we had been given?

'I love a cup of Earl Grey, don't you?'

'Yes, but this is Darjeeling.'

'It's Earl Grey.'

'It is Dar-bloody-jeeling!'

Or was it simply an age-old tribalism?

'Of course, I am a coffee man myself.'

'Well I am a tea man.'

'No, no, you can't beat coffee.'

'Tea.'

'Coffee.'

'Tea.'

The truth is I do like coffee. But I also like tea. But which is best? Of course, there was only one way to find out. *Fight!* At which point I clouted John with my tripod.

Give it to him

And where did this assault take place? The fact that I had a tripod with me suggests that it was either a photographic or birding context. Maybe he was a guest on one of my shows, or I was a guest on his. I have certainly met and several times worked with John Craven, but I like and admire him, and I hope the feeling is mutual. Indeed, I would go so far as to say that were we to find ourselves when there was only one cup of tea left in the pot, I would gladly give it to him. Or does he prefer coffee?

Slapper

The tripod myth baffles me, unless it stems from the time the two of us were broadcasting from the Chelsea Flower Show, and I got a bit too playful with a bunch of gladioli while impersonating Dame Edna. I admit I started it, but John did retaliate by giving me a damn good thrashing with a Red Hot Poker.

By the way, this paragraph is as fictitious as the stuff on the website.

And another thing

While I'm at it, I have been implicated in several other urban myths (why just urban? Does that imply that people are less gullible in rural areas? Of course those townies will believe anything!).

I have read and been told several times that I live in Norfolk. I don't, nor ever have, but I can understand why people think I should. In fact, I have lived in London for 50 years. Before that I lived in Birmingham, and it is true that I went to King Edward's School, Birmingham. However, the story that on the day of the Queen's visit I moved all the bollards and rerouted the whole royal convoy through the school is not true. But I wish it was. Finally, it is a matter of rock and roll legend that The Goodies once beat up John Peel after he gave the 'Funky Gibbon' a bad review. No, they didn't.

Index

acid rain 146
acorns 170–171
Addis Ababa, Ethiopia 44–47
airborne birding 85–87
airport birding 42–44
Albatross, Waved 56
animal behaviour 70–72
animal costume 156–157
Ankylosaurus 95–96
anthropomorphism 65–67
Arizona, USA 153–154
artwork 101–105
Attenborough, Sir David 54–55, 98–99, 185
  Life in the Undergrowth 207
Auk, Little 93–94
Australia 106–108, 112
Autumnwatch 10–11

bad weather birding 38–41
Badger cull 137, 187
barbtails 153
BBC 10, 11, 64, 95, 111, 121, 150, 182, 183
  Natural History Unit 62, 86
  Radio 126–128
  TV Centre 204
BBC Wildlife magazine 167, 183, 208
Bee-eater 36
bees 201–202, 207
Bentbill, Northern 28
Bill Oddie Goes Wild 115–116
binoculars 27–28, 208
bipolar disorder 10–11, 210
bird calls 123–125
bird food 171–172, 190–191
Bird in the Nest 62–64
birding technology 27–29
Birding with Bill Oddie 88
Birdline 28
birdwatching 75–76
Blackbird 68–69, 171, 172, 188, 191, 197
blogging 167
  News of the Wild 169–170
Bluethroat 152
Booby, Blue-footed 55
Borneo 147–148
BP 196
Braer 146
Breakaway 126–128
Brindled Beauty 115
Brooke-Taylor, Tim 9, 156, 217
Buffalo, African 44
Buglife 202
Bumblebee, Tree 202
  White-tailed 201
Bunting, Reed 124
  Snow 122
Burton, John 195
Busby, John 103
Buzzard 17, 138–139, 161
  persecution 139–140, 187
buzzards 35

Cambridge Sewage Farm 112
Canada 49–50
Capercaillie 88–91
Chaffinch 124, 172
Channel 5 182
children 22–24
Chlorophonia, Blue-crowned 28
Cleese, John 55, 217
Colby, Flick 205
Compassion in World Farming 11
Corbett National Park, India 130
Cormorant 143
  Flightless 56, 121
Costa Rica 30, 126–128
Country Innovation 208
Countryside Alliance 187
Coyote 153
cranes 89, 92–93
Craven, John 217, 218, 219
Creationism 98–99
Crow, House 43
Cuckoo 34, 124
Curlew 36, 119

Daisy 184
darters 118
Darwin's finches 53, 54, 56
DDT 161
decoys 16
Deer, Spotted 131
Defra 137, 140
depression 209–210
Dinosaur Valley, Texas 97–98
dolphins 48, 69, 142
Dunnock 17, 171

Eagle, Crowned 87
  Golden 138, 161
  White-tailed 52, 161
eagles 35

egg collecting 23
Egret, Cattle 42
egrets 112
Elephant, Pygmy 148
elephants 12, 129–131
emerald mining 195, 196
Ennion, Eric 81–82, 103, 105
environmental destruction 146–148, 163–165, 195–196
Ethiopia 44–47

fake birds 16–17, 43
Falcon, Eleanora's 104–105
  Peregrine 17, 35, 104, 139, 150, 161
falcons 35, 150
farmers 135–137
Farne Islands, UK 55, 80, 81–82, 103
field notes 101
filming wildlife 61–64
Finch, Vampire 56
  Vegetarian 56
  Woodpecker 56
Firewood-gatherer 153
first morning birding 152–154
Flycatcher, Red-breasted 152
  Scissor-tailed 96
  Yellow-bellied 29
flycatchers 154
foliage-gleaners 153
Fowl, Jungle 130
Fox 20–21, 180–181, 187–188, 190
frigatebirds 55
frogspawn 174, 177–179, 181, 187–88
fungi 207–208

Galápagos 30, 53–57, 121
gamekeeping 159–160, 161
gannets 42
garden lists 17–18
garden ornaments 15–17
  fake predators 16–17
  mirrors 17
  miscellaneous 179–180
garden ponds 16, 174–175
  frogspawn 174, 177–179, 181, 187–188
  installing 175–176
  replacing 176–177
Garden, Graeme 9, 156
geese 35, 143, 183
gibbons 148, 198, 199, 201
Global Witness 11
Gnome Corner 16, 20
Goldcrest 124
Goldfinch 92, 93, 124, 171
golf courses 117–119
Goodies, The 9, 86, 155–156, 219
Goose, Canada 70–71
Grahame, Kenneth *The Wind in the Willows* 66
Great Gorilla Run 157–158
Great Storm 1987 37–38, 147
Greenfinch 172
Ground Finch, Large 56
  Medium 56
  Small 56
Grouse, Black 66
Guan, Highland 33
Guatemala 28–29
  volcano birding 30–33
Guillemot 81
  Bridled 39
Gull, Black-headed 55, 128
  Great Black-backed 82
  Lava 30, 55, 121
  Sabine's 38
  Swallow-tailed 55
gulls 42, 52, 112

Hamilton James, Charlie 62
Harrier, Marsh 161
harriers 43
Hawk, Harris 149–151
Hawkmoth, Death's-head 115
hearing 123–125
Hebrew Character 115
Heron, Lava 30, 121
herons 16, 112, 118
herring 51–52
hippos 153
Hobby 17, 104, 161
Holden, Peter 62, 63
Honey-buzzard 17
Hoopoe 36, 102
hot-air ballooning 85–86
Houses of Parliament 185–186
Humane Society 11
hurricanes 38–41, 78
HUTAN 148

I'm a Celebrity... Get Me Out of Here 206–207
Ibis, Wattled 46
ibises 118
Iceland 48–49, 51–52, 142–145
Iguana, Marine 55
India 129–131, 153
insects 207
Inspired by Nature 11

International Fund for Animal Welfare (IFAW) 11, 49, 144
 Song of the Whale 144–145
International Primate Protection League (IPPL) 199, 201

Jackdaw 63, 139
Jackrabbit 150
Jay 20, 171, 179, 188, 191, 212
Junco, Volcano 30, 121
Jungle Book, The 66–67

Kearton, Cherry 62
Kenfig National Nature Reserve, UK 119
Kenya 86–87, 153
Kestrel 17, 35, 63, 104, 139
King, Simon 62, 63, 185, 118
Kingfisher 63, 64
kingfishers 118
Kite, Red 17, 139, 161
Krause, Bernie *The Great Animal Orchestra* 69

landowners 160–161
Lapwing 23, 197
larks 43, 117
laser pens 29
Leaftosser, Scaly-throated 28
League Against Cruel Sports 11, 140, 186
LEAP 148
Lion 44, 197
lists 42
Lonesome George 55

Mabey, Richard 112
Magpie 139, 179, 188, 191, 212
 Azure-winged 36, 102
Maiden's Blush 115
meadowlarks 96
Merlin 104
Mitchell, Joni 164
mockingbirds 56
Monkey, Proboscis 148
Monks' House Bird Observatory 81, 103
Moray Firth, Scotland 51
mosquitoes 207
Moss, Stephen 88, 89–91, 183
moths 114–116, 212–213
Mouse, Wood 191–193
Mullarney, Killian 105
Myna, Common 43

National Trust 184
 Natural Childhood 22–24
Natural England 184
Natural History Museum, London 185
Nature Deficit Disorder 22, 24
New Jersey, USA 38–41
Newsnight 11
NFU 137
NGOs 11–12, 137, 148, 160, 167, 185, 210
Nightingale 106
Nightjar, Red-necked 102
noise pollution 51
Norfolk, UK 146, 184, 219
Norfolk Ornithologists' Association 184
Norfolk Wildlife Trust 184
Northumberland, UK 196

Oddie in Paradise 111–112
Oddie, Bonnie 206
 Kate 206
 Laura 153, 154, 172–173, 206
 Rosie 114, 153, 154, 184, 206
Old Lady 114, 115
open-cast coal-mining 196
Orangutan 12, 148
Orca 49–52
Orchid, Lizard 118
Oriole, Golden 102
orioles 154
ornithology 76
Oropendola, Montezuma 128
Osprey 17, 35, 161
Owl, Barn 71
 Little 36
 Snowy 43

palm oil production 147, 148
Pan's People 20
Papua New Guinea 111–112, 153
Parakeet, Monk 19, 35, 203
 culling 203
Parakeet, Ring-necked 19, 20, 150, 202–203, 212
parakeets 139, 203
Parliament Hill, London 124
Parrot, Orange-bellied 112
 Seychelles Black 121
Patagonia, Argentina 50
pâté de foie gras 143, 183
Peacock 130
Peel, John 219
penguins 50
perches 34–36
Perry Oaks Sewage Farm 112–113
Petrel, Leach's 38
Pheasant 139, 140, 141
Pigeon, White-collared 46

pigeons 150, 188
Pipit, Meadow 34, 124–125
pipits 43, 117, 197
Plover, Oriental 43–44
  Ringed 113
Poás Volcano, Costa Rica 30
poles 35, 36
porpoises 48, 142
power lines 35, 36
Pratincole, Black-winged 136
Prominent, Three-humped 116
Pug, Triple-spotted 116
Puffin 80–84, 142, 143
Purple Treble-bar 115
pylons 35–36

Quail 106
  Gambel's 154
Quaker, Common 115

Rabbit 141
Radio Times 95
raptors 35, 104, 139, 150, 154, 160
  persecution 159–161
rats 17, 20
Razorbill 81
Reading Football Club 167
reedhaunters 153
Richardson, R. A. 103, 105
Roadrunner 67, 96–97, 154
Robin 17, 171, 172, 179, 188, 191, 211–212, 213–214
  Sooty 121
Roller 36
Rose, Chris 105
Royal St George's Golf Course, Kent 118
RSPB 11, 62, 92, 93, 136, 137, 140, 160, 184, 198
  Big Garden Birdwatch 17–18, 76, 197
  Osprey Visitor Centre, Loch Garten 88–89
RSPCA 11, 93, 186
Ruff 113

Sandpiper, Buff-breasted 78–79
  Semipalmated 118
Scalloped Hook-tip 115
Scilly Isles, UK 53, 78, 104
Scrub-bird, Noisy 106–108
scrumping 23
Seal, Elephant 50
  Grey 55
sealions 53, 55, 56
seals 50–51
Seedeater, Brown-rumped 46
sewage farms 111–113
Seychelles 54, 121
Shag 143
Shearwater, Sooty 39
shearwaters 38
Shell 196
Sheppard, Andy 68–69
Shetland, UK 34, 54, 146
Shrike, Southern Grey 36
  Woodchat 36, 102, 103
shrikes 35, 154
Sielmann, Heinz 62
Skomer, UK 83–84
Skua, Great 82
Skylark 23, 197
Snipe, Painted 152
Snout, Pinion-streaked 115
songbirds 143
sound recording 68–69
South Carolina, USA 199–201
Sparrow, House 43, 45, 54, 71, 128, 171, 197
  Lark 96, 97
Sparrowhawk 17, 71–72, 139
sparrows 154
Spey Valley, Scotland 88
Springwatch 10, 71, 184
Squirrel, Grey 170, 171, 172
  Red 170
SSSIs 136
Starling 72, 197
Stork, White 84
storks 35
Strictly Come Dancing 204–206
stringing 102
Surtsey, Iceland 121
Swallow 35, 121–122, 124
  Red-rumped 36
swallows 43
Swan, Mute 36, 70–71
swans 35
Swarovski 208
Sweden 89–91
swifts 43

Tanzania 44
telegraph wires 34–35
telescopes 27
Tern, Bridled 39, 41
  Least 118
  Sooty 39, 41
terns 112
Texas, USA 96–100
Thrush, Song 171, 172, 191
thrushes 117

Tiger 129–131
Tit, Blue 17, 71, 94, 171, 181, 191
  Great 17, 63, 171, 188
Tortoise, Giant 54, 55
Toucan, Chestnut-mandibled 128
Treecreeper 123–124
trespassing 23
Triceratops 95
tropicbirds 42
True Lover's Knot 115
Truth About Killer Dinosaurs, The 95–100, 151
Turkey, Wild 99
tweeting 184
twitching 76
  vagrants 77–79
Tyrannosaurus rex 96

untruths 217–219

vagrants 77–79, 93
velociraptors 99, 150–151
volcano birding 30–33, 120–122
Vulture, Turkey 96

waders 112, 113, 118
Wagtail, White 78
Wales, UK 196
Walt Disney World, Florida 118
Warbler, Cetti's 106, 107
  Grasshopper 106, 124
  Tennessee 128
warblers 32, 117, 118, 124, 154
wasps 20
watermelons 203
Watson, Chris 69
waxbills 102
weather 37–38, 189–191
  hurricanes 38–41, 78
  tornadoes 199–200
weavers 35
Werribee Water Treatment Works, Australia 112
Whale, Killer 50–51
  Minke 49, 143–144, 145
whales 12, 48, 142
whale-watching 48–52, 142–145
whaling 142–145
Wheatear 34
wheatears 78, 117
Whimbrel 119
  Little 119
Wild in Your Garden 71–72
Wildfowl & Wetlands Trust 137
wildfowlers 16
Wildlife Trusts 11, 137, 140, 163, 184, 185, 202
Wildscreen 65
Willow Beauty 213
Wisbech Sewage Farm, UK 111, 112
woodcreeper 32
Woodpecker, Great Spotted 61–64, 94, 212
woodpeckers 35, 118, 154
Woodpigeon 17, 20
World Land Trust 11, 147–148, 195, 196
Wren 17, 69, 188, 197
  Cactus 153–154
writing 12, 183

Yellowhammer 23

Zambia 194–195, 196–197